Multi-Agent Energy Systems Simulation

Multi-Agent Energy Systems Simulation

Editors

Tiago Pinto
João Soares
Fernando Lezama

MDPI • Basel • Beijing • Wuhan • Barcelona • Belgrade • Manchester • Tokyo • Cluj • Tianjin

Editors
Tiago Pinto
BISITE Research Group,
University of Salamanca
Spain

João Soares
GECAD—Research Group on
Intelligent Engineering and
Computing for Advanced
Innovation and Development,
Institute of Engineering,
Polytechnic of Porto (ISEP/IPP)
Portugal

Fernando Lezama
GECAD—Research Group on
Intelligent Engineering and
Computing for Advanced
Innovation and Development,
Institute of Engineering,
Polytechnic of Porto (ISEP/IPP)
Portugal

Editorial Office
MDPI
St. Alban-Anlage 66
4052 Basel, Switzerland

This is a reprint of articles from the Special Issue published online in the open access journal *Energies* (ISSN 1996-1073) (available at: https://www.mdpi.com/journal/energies/special_issues/Multi_Agent_Energy_Systems).

For citation purposes, cite each article independently as indicated on the article page online and as indicated below:

LastName, A.A.; LastName, B.B.; LastName, C.C. Article Title. *Journal Name* **Year**, *Article Number*, Page Range.

ISBN 978-3-03943-649-1 (Hbk)
ISBN 978-3-03943-650-7 (PDF)

Contents

About the Editors

Tiago Pinto was awarded his PhD in 2016 and is currently Researcher at the GECAD research group, ISEP/IPP. He has participated in more than 20 research projects, including leadership of H2020 MSCA-IF. He has published over 200 scientific papers in his main areas of research interest, which include the application of artificial intelligence (AI) in several problem domains, especially in the fields of power and energy systems, electricity markets, and smart grids. His main expertise lays in the fields of adaptive machine learning and automated negotiation, including multi-agent systems, machine learning algorithms, knowledge-based systems, game theory, artificial neural networks, genetic algorithms, particle swarm intelligence, and data mining. Tiago Pinto has been awarded multiple prizes and awards for works developed in these fields, e.g., APPIA (Portuguese Association for AI) best PhD thesis in AI and REN (Portuguese Electrical Network Operator) 2nd best MSc thesis.

João Soares has a BSc in Computer Science (2008) and a master's degree in Electrical Engineering (2011) from Polytechnic of Porto, Portugal. He attained his PhD degree in Electrical and Computer Engineering (2017) at UTAD university. He is currently Researcher at GECAD—Research Group on Intelligent Engineering and Computing for Advanced Innovation and Development in the School of Engineering of the Polytechnic of Porto and has recently been an invited professor at Ecole Centrale De Lille in the L2EP. He coordinates projects in the field of smart grids and smart buildings with application of computational intelligence techniques. His research interests include optimization of power and energy systems, including heuristic, hybrid, and classical optimization. He has published more than 140 publications in this field and his works have been cited over 2500 times (H-index of 25 in Google Scholar).

Fernando Lezama received his MSc degree (with Honors) in Electronic Engineering—Telecommunication (2011), and the PhD in ICT (2014) from the Monterrey Institute of Technology and Higher Education (ITESM), Mexico. He was also a postdoctoral researcher at the National Institute of Astrophysics, Optics, and Electronics—INAOE (2015–2017), Mexico, where he worked in the development of intelligent systems for optimization in smart grids. Since 2017, he has served as Researcher at GECAD, Polytechnic of Porto, where he contributes to the application of computational intelligence (CI) in the energy domain. Dr. Lezama has been part of the National System of Researchers of Mexico since 2016, Chair of the IEEE CIS TF 3 on CI in the Energy Domain, and has been involved in the organization of special sessions, workshops, and competitions (at IEEE WCCI, IEEE CEC and ACM GECCO) to promote the use of CI to solve complex problems in the energy domain. He is also part of the National System of Researchers (I-Level) of Mexico. He has been author and co-author of more than 50 academic papers published in top-tier journals and conferences in telecommunications, computational intelligence, and power systems. His research interests include computational intelligence, evolutionary computation, network planning, and optimization of smart grids and optical networks.

Preface to "Multi-Agent Energy Systems Simulation"

The synergy between artificial intelligence and power and energy systems is providing promising solutions to deal with the increasing complexity of the energy sector. Multi-agent systems, in particular, are widely used to simulate complex problems in the power and energy domain as they enable modeling of dynamic environments and studying the interactions between the involved players. Multi-agent systems are suitable for dealing not only with problems related to the upper levels of the system, such as the transmission grid and wholesale electricity markets, but also to address challenges associated with the management of distributed generation, renewables, large-scale integration of electric vehicles, and consumption flexibility. Agent-based approaches are also being increasingly used for control and to combine simulation and emulation by enabling modeling of the details of buildings' electrical devices, microgrids, and smart grid components.

This book discusses and highlights the latest advances and trends in multi-agent energy systems simulation through a collection of 8 research papers. The addressed application topics include the design, modeling, and simulation of electricity markets operation, the management and scheduling of energy resources, the definition of dynamic energy tariffs for consumption and electrical vehicles charging, the large-scale integration of variable renewable energy sources, and mitigation of the associated power network issues.

"The Application of Ontologies and Multi-Agent Systems in the Energy Sector: A Scoping Review" provides a scoping review of the existing literature on ontology for multi-agent systems in the energy domain, and maps the key concepts underpinning these research areas. Furthermore, this paper provides a recommendation list for the ontology-driven multi-agent systems development.

"Variable Renewable Energy and Market Design: New Market Products and a Real-World Study" addresses the topic of market design to accommodate large-scale integration of variable renewable energy. A new bilateral energy contract type is proposed along with two new marketplaces that can contribute to reducing the imbalances resulting from variable renewable energy producers are introduced.

"Dynamic Tariff for Day-Ahead Congestion Management in Agent-BVased LV Distribution Networks" advances the research made in the area of congestion management in low-voltage networks. The paper tackles these challenges by iterative chances in prices (dynamic tariffs). An agent-based system is able to demonstrate a reduction of 82% in congestion using an IEEE European LV test feeder without loss of power quality in the grid.

"An Optimal Energy Management System for Real-Time Operation of Multiagent-Based Microgrids Using a T-Cell Algorithm" proposes the design and implementation of a real-time energy management system based on a multi-agent systems approach. The fast converging T-cell algorithm is applied to minimize the operational cost of a microgrid and maximize the real-time response in grid-connected microgrid mode.

"Reactive Power Management Considering Stochastic Optimization under the Portuguese Reactive Power Policy Applied to DER in Distribution Networks" provides a stochastic agent framework to improve the reactive power management by taking advantage of the full capabilities of the distributed energy resources and by reducing the injection of reactive power by the transmission system operator in the distribution network and, therefore, reducing losses. The uncertainty of renewables is considered in the proposed sequential alternative current optimal power flow.

"Using Agent-Based Customer Modeling for the Evaluation of EV Charging Systems" takes further steps in the modeling of customer behavior when it comes to charging electrical vehicles. Previous works usually adopted a stochastic approach with few details since little information is available. This work uses more detailed customer model employing a multi-agent simulation framework in order to investigate how a customer behavior that responds to external factors (like weather) or historical data (like satisfaction in past charging sessions) impacts the essential key performance indicators of the charging system. Results show that the MAS system can produce quantitative and qualitative differences when small changes are tested in the customer behavior.

"Wireless Sensor Network Energy Model and Its Use in the Optimization of Routing Protocols" presents an energy model that estimates the energy consumption at each node of a network, taking into account the functions of sensors transmitting data. Therefore, the model considers a given routing protocol allowing the comparison and assessment of different performance metrics from an energy standpoint. The model was validated on a real proof-of-concept implementation using system-on-chip equipment. The proposed model achieved 97% accuracy compared to the actual performance of a network, which reflects its effectiveness in comparing communication protocols in WSNs.

"Smart Campus: An Experimental Performance Comparison of Collaborative and Cooperative Schemes for Wireless Sensor Network" objectively defines a set of performance metrics to compare different IoT communication protocols used in wireless sensor networks. A real wireless sensor network is placed on a university campus to compare the performance of some of the most popular protocols, i.e., Zigbee, LoRa, Bluetooth, and WiFi. Since energy consumption is a crucial aspect of the wireless sensor network (due to battery-powered sensors), particular focus is given to the metrics related to energy efficiency. The defined performance metrics and methodology become a suitable tool in contrasting low-consumption wireless technologies applied to IoT that can be used to implement multi-agent systems.

The complementarity and broad scope of this collection of papers offers a relevant perspective of many challenges arising in power and energy systems, and how multi-agent simulation approaches are contributing to overcoming such challenges.

We thank all the authors and reviewers who have significantly contributed to the high quality of the papers included in this collection. We also express our gratitude to the editorial team of MDPI and *Energies* for all the support during the entire project.

Tiago Pinto, João Soares, Fernando Lezama
Editors

Review

The Application of Ontologies in Multi-Agent Systems in the Energy Sector: A Scoping Review

Zheng Ma [1], Mette Jessen Schultz [2], Kristoffer Christensen [3], Magnus Værbak [3], Yves Demazeau [4] and Bo Nørregaard Jørgensen [3,*]

1   Center for Health Informatics, University of Southern Denmark, Campusvej 55, 5230 Odense, Denmark
2   Danish Energy Agency, Niels Bohrs Vej 8D, 6700 Esbjerg, Denmark
3   Center for Energy Informatics, University of Southern Denmark, Campusvej 55, 5230 Odense, Denmark
4   Laboratoire d'Informatique de Grenoble, Centre National de la Recherche Scientifique, 700 avenue Centrale, 38000 Grenoble, France
*   Correspondence: bnj@mmmi.sdu.dk

Received: 25 June 2019; Accepted: 15 August 2019; Published: 20 August 2019

Abstract: Multi-agent systems are well-known for their expressiveness to explore interactions and knowledge representation in complex systems. Multi-agent systems have been applied in the energy domain since the 1990s. As more applications of multi-agent systems in the energy domain for advanced functions, the interoperability raises challenge raises to an increasing requirement for data and information exchange between systems. Therefore, the application of ontology in multi-agent systems needs to be emphasized and a systematic approach for the application needs to be developed. This study aims to investigate literature on the application of ontology in multi-agent systems within the energy domain and map the key concepts underpinning these research areas. A scoping review of the existing literature on ontology for multi-agent systems in the energy domain is conducted. This paper presents an overview of the application of multi-agent systems (MAS) and ontologies in the energy domain with five aspects of the definition of agent and MAS; MAS applied in the energy domain, defined ontologies in the energy domain, MAS design methodology, and architectures, and the application of ontology in the MAS development. Furthermore, this paper provides a recommendation list for the ontology-driven multi-agent system development with the aspects of 1) ontology development process in MAS design, 2) detail design process and realization of ontology-driven MAS development, 3) open standard implementation and adoption, 4) inter-domain MAS development, and 5) agent listing approach.

Keywords: multi-agent system; ontology; energy sector; scoping review

---

1. Introduction

The energy sector is facing a new paradigm shift following the large-scale integration of renewable energy sources (RES) [1]. The significant use of fossil resources is one of the major concerns of today's society. Climate changes, environmental impacts, and the scarcity of resources have led to the need for RES. RES reduce greenhouse gas emission while contributing to an increase in life quality and sustainable development [2]. The inclusion of RES is a highly complex task. The demand and supply need to be balanced due to the unpredictable behavior of RES. This influences not only the electricity system but also heating and cooling systems due to the considerable linkage between subdomains.

In order to solve these problems, multiple stakeholders need to work together and provide solutions. Models of such solutions are essential to explore the interactions between consumption, production, and transportation as well as economic, environmental and technical phenomena. Multi-agent systems (MAS) can contribute to explore and develop such solutions since MAS can simulate how multiple stakeholders work, interact, and influence each other. The MAS simulations make it possible to simulate systems which consist of agents with different or conflicting objectives.

Agents often collaborate towards a specific goal and need to communicate and share results. Different languages and vocabularies are domain-specific, and often cause problems for the agents in a system. It requires a common language to ensure that messages are interpreted correctly between agents [3]. Therefore, ontology can be applied to establish effective communication between agents. Ontology can specify terms that are used for communication within a specific context and enable agents to make declarations or ask queries that are understood by all other agents in the system [4]. It is an important tool for the development of an intelligent multi-agent energy system, e.g., for the knowledge sharing and knowledge reuse [5].

As more applications of multi-agent systems in the energy domain for advanced functions, the interoperability challenge raises due to an increasing requirement for data and information exchange between systems. Meanwhile, the energy system is strongly connected with other domains. Therefore, the application of ontology in multi-agent systems needs to be emphasized and a systematic approach for the application needs to be developed.

Although some review papers have investigated agent-based modeling and tools for the electricity domain (e.g., [6]), very few studies have investigated the MAS design and the applications of ontology in MAS for the energy domain. Moreover, many studies focus on specific subdomains and how to solve one specific problem. Hence, investigation and analysis of more complex systems and problems, integration of subdomains, including different agents and ontologies, is needed. Meanwhile, it is important to highlight the relevant literature and map the key concepts underpinning the research area [7]. The scoping review can provide the means that identify, characterize, and summarise existing literature regarding the state of research activities. Moreover, the review result can identify gaps in the literature.

This paper conducts a scoping review to investigate the existing studies on the application of ontologies in the MAS for the energy domain. Based on the results of the literature analysis, this paper proposes a recommendation list for the ontology-driven MAS development for the energy domain. This recommendation aims to address certain aspects that are missing in the literature or need more emphasis in future work.

The paper is organized as follows: Section 2 describes the methodology and the research process. Section 3 presents the literature analysis results, and Section 4 discusses the findings followed by Section 5 that concludes. The conclusion section also states the recommendation for future work and the limitations of this study.

2. Method

The study is designed to compile the relevant contributions from previous publications and to analyze their results in relation to multi-agent modeling design for the energy domain. This study firstly conducts a literature search of ontologies and multi-agent systems for the energy domain. The literature search was performed during the first quarter of 2019. To retrieve the relevant articles for this literature study, four online databases are selected that are relevant in the fields of energy, and MAS and ontologies: ACM digital library, IEEE Xplore, Web of Science, ScienceDirect. The review covers books, conference proceedings, academic journal articles, research articles, and review articles. Other forms of publications, such as newspapers, posters, etc., were not considered since their publication forms are not for scientific research purposes. There was no limitation on the publication years for the literature search.

The data collection was divided into three rounds with relevant keywords. The keyword search was only applied to titles due to a large number of the literature in the fields and the concerns of the relevance in the selected domains. The first round focused on the multi-agent systems in the energy domain. To avoid excluding any relevant study, the search strings were:

('multi-agent' OR 'multiagent') AND ('energy' OR 'electricity' OR 'heating' OR 'grid' OR 'electric' OR 'power' OR 'wind')

The strings, in the first round, resulted in 1433 publications. The result from each database is shown in Table 1. All these 1433 publications were imported to the reference management software-Endnote (https://endnote.com/).

Table 1. Results in the first round search.

Database	Result
Web of Science	355
IEEE	822
ScienceDirect	58
ACM	198
Total	1433

To dismiss the duplicated publications, i.e., articles which were obtained through multiple databases or strings, 856 articles were removed by this criterion. The remaining 577 articles were selected for further analysis. This study searched the remaining articles with 'ontology' OR 'ontologies' in titles, abstracts, and keywords, and resulted in 24 articles with full-text.

Based on the text mining in the analysis software NVivo (https://www.qsrinternational.com/nvivo/home) and careful review, the 24 articles were separated into six sub-domains (shown in Table 2). Majority of the selected articles only address one sub-domain, and one article [8] addresses three sub-domains (energy management, microgrid, and buildings), and another article [9] addresses two sub-domains (power system and microgrid). The publications show that the application of ontologies in the field of MAS for the energy domain was mainly conducted after the year 2004, with focus on the sub-domain of grid control between 2004 to 2014, and expanded into the sub-domain of electricity market since 2014. A list of the 24 articles in the Appendix A shows the focused aspects in the energy domain, ontology, and MAS design.

Table 2. Six addressed sub-domains by the selected articles.

Grid Control	Power System	Energy Management System	Microgrid	Buildings/Demand Side	Electricity Market
8	3	2	3	5	6

3. Results

This study reviews and analyses the selected 24 articles to investigate the current research on the application of ontologies in MAS for the energy domain, and the main discussion in the 24 articles can be divided into five categories: 1) definition of agent MAS, 2) MAS applied in energy domains 3) defined ontologies in the energy domain, 4) MAS Design and architectures, and 5) Ontology in the MAS development.

3.1. Definition of Agent and MAS

3.1.1. Agent and Agent-Based Modeling

An agent is defined as an entity that reacts to changes in its environment through a reasoning process [10]. The attributes of an agent are autonomy, sociability, reactivity, pro-activeness, adaptiveness, interactivity, rationality, and interactivity, etc. [11]. Russell [12] defines an intelligent agent as an autonomous entity which has the following properties:

- It has the ability to communicate and interact with its environment;
- It is able to perceive the (local) environment;

- It is guided by basic objectives;
- It has feedback behavior.

An agent structure shows: (1) a set of modules that the agent is decomposed in to, (2) the interaction between these modules and the environment and other agents (shown in Figure 1), and generally, there are three types of agent structures: deliberative architecture, reactive architecture and hybrid architecture [13].

Agent-based modeling is a model of a system with the description of agents and agents' interactions [14]. Agent-based modeling usually models part of the system rather than a whole system due to the complexity of the system.

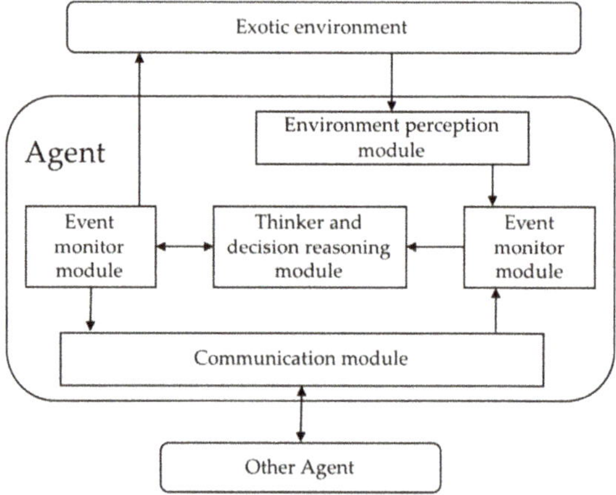

Figure 1. Agent structure [13].

3.1.2. Multi-Agent Systems

Multi-agent System (MAS) is a complex system that is composed by more than one distributed agents and these agents communicate to deal with problems which usually can't be solved by a single agent [14,15]. According to [16], a MAS is characterized by:

- Large numbers of actors are able to interact, in competition or in cooperation;
- Local agents focusing on local interests and negotiating with more global agents;
- Implementation of distributed decision making, through negotiation processes between different local or global agents;
- Communication between actors is minimized to generic information exchange between agents: only the information necessary for their functioning is sent between agents.

MAS is based on the divide-and-conquer mechanism [17]. In a MAS, each agent has limited knowledge about its environment, and work individually towards a certain goal based on their local knowledge and their behavioral algorithms and interact in a cooperative or competitive manner with other agents [18].

The idea of using MAS is to divide a complex system into smaller and more related objectives and construct agents for these sub-objectives [17]. MAS can simulate and control large complex decentralized systems that can cope with the dynamics of the system, reduce the complexity, and increase flexibility [19]. One of the most important benefits of MAS is its fault tolerance, based on multiple agents can provide the same services [17].

3.2. MAS Applied in Energy Domains

The energy sector is becoming more complex and consists of multiple hybrid systems, which includes various interactions and amounts of knowledge. MAS is being studied in many areas of power engineering including diagnostics, condition monitoring, power system restoration, market simulation, network control and automation, and hierarchical decision making, as smart grid (SG) and microgrids (MG) [18,20]. The development of simulation platforms based on MAS is increasing as a good option to simulate real systems in which stakeholders have different and often conflicting objectives [21].

3.2.1. MAS for Grid Control

According to [18], research on using MAS in power engineering mainly focuses on distributed control architectures and simulation. MAS is a decentralized scheme that utilizes distributed controllers for energy management and optimization, and it is an alternative approach for smart system optimizers (SSOs) implementation within a typically integrated energy system (IESs) [22]. MAS is an obvious and promising choice for the smart grid control system because MASs can overcome the threat of SPOFs (single-point-of-failure) due to their distributed characteristic [23]. Meanwhile, Considering the agent properties, the variety of components used in power transformer and the huge amounts of data involved, MAS provides the best possible choice for the purpose of monitoring, automating, controlling and diagnosing the power transformer components [24]. MAS has proven to be suitable for addressing the demands of SGs both theoretically and practically [25].

Most of the research work in this area have focused on hierarchical control, optimization, and power restoration using MAS. For instance, [21] proposes a MAS-based optimal energy management solution for the optimization problem of the interactive operation of generation units and DR [26]. Similarly, introduces a decentralized agent-based approach for optimal residential demand planning [27]. A MAS is used in [28] to restore power in case of failure, and [29] introduces a flexible and versatile MAS for fault isolation and power restoration. Meanwhile, [30] presents a MAS automated management and analysis of SCADA and Digital Fault Recorder Data. Furthermore, a multi-agent system is used to control the voltage of the power system with co-ordination in [31].

Other distributed MAS-based solutions to grid control are also presented microgrids, islanded microgrids, and multiple microgrids [8]. The applications of MAS in a microgrid is similar to the smart grid control, e.g., Microgrid control, optimal energy exchange, and multi-level management, but also link to buildings or demand-side management. For instance, [32] presents a MAS for Microgrid control and a classical distributed algorithm. [33] proposes a MAS microgrid system for optimal energy exchange between the production units of the Microgrid and local loads. based on MAS, [34] proposes an Intelligent Distributed Autonomous Power System (IDAPS) to increase the reliability of the critical loads. [35] proposes a multi-level management and control scheme for microgrid systems taking into account the interaction among agents at different levels. [36] presents a consumption scheduling framework in small residential areas.

3.2.2. MAS for Electricity Markets

MAS of the electricity markets concern market players and markets modeling, strategic bidding and decision support [37]. Multi-agent-based simulation of the electricity markets usually combines with artificial intelligence techniques and game theories and is not only simulation platforms but also provides opportunities for the scenario comparison, future evolution study and sensitive analysis [38].

Several studies have applied MASs to model and simulate electricity markets [14]. For instance, Li et al. [39] discuss the potential for developing Open Source Software (OSS) for power market research. The Agent-based Modelling of Electricity Systems (AMES) is an agent-based OSS laboratory, specifically designed for the experimental study of reconstructed wholesale power markets. The AMES simulation includes an independent system operator, load-serving entities, and generation companies distributed across the transmission grid.

Another electricity market model is the Electricity Market Complex Adaptive System (EMCAS) model [40] utilized by Koritarov [1]. The model is used to capture and investigate the complex interactions between the physical infrastructures (generation, transmission, and distribution) and the economic behavior of market participants [41]. Furthermore, the model applies an agent-based approach where agents' strategies are based on learning and adaption. This approach enables simulations in different time periods, from real-time to decades including both pools and bilateral contract markets. This approach also makes it possible to see the evolution of an electricity market over time and stakeholders' reaction towards changes in economy, finance, and regulation. The study describes two methods of how the agents learn: observation-based and exploration-based learning. In observation-based learning, the learning process is based on a structured process of past market performance evaluation, future market status prediction, and investigation of other agents' actions. Agents decide either to keep or adjust their current market strategy or use a new strategy. Agents based on exploration-based learning explore new market strategies, and these strategies are simulated in a simulation tool. The results are observed, and the strategies are either accepted or rejected based on the results and the agents' goals.

Praca et al. [42] develop the Multi-Agent Simulation of Competitive Electricity Markets (MASCEM) [43]. The model is developed to study the behavior and evolution of an electricity market. The MASCEM is a modeling and simulation tool aiming to study the operation of complex and competitive electricity markets [44]. The agents in the system represent the market entities, such as generators and customers. The MASCEM allows agents to establish their own decision rules and adapt their strategies as the simulation progresses based on previous events. As a decision-supporting tool, the simulator includes different possibilities regarding electricity market negotiations [45,46]. The MASCEM is a flexible tool which makes it easy for users to define models including strategies, types of agents and market types. For example, this flexibility is utilized by Santos et al. [3,47,48] for modeling and simulating the EPEX (central European electricity market) and Nord Pool spot market (Scandinavian electricity market). The MASCEM can also be used for modeling and simulation of other electricity markets such as MIBEL (the Iberian electricity market), GME (the Italian electricity market), and even markets outside Europe [48].

3.2.3. MAS for Demand-Side and Building Systems

MAS provides a flexible and reliable solution to manage and optimal loads at demand-side with the consideration of energy cost minimization and user's comfort maximizations [49,50]. MAS has been applied in automated building management systems (BMS) for energy-related building research [16,51–53].

The automated BMS research in energy-related building systems mainly focuses on control mechanisms of building loads and investigate possibilities and potentials of energy efficiency and flexibility in buildings [54,55], and especially much equipment in buildings can be controlled and deliver demand flexibility, e.g., lighting and HVAC, and can respond to the grid signals [56]. Although complex control systems are important in building systems, these processes need to be optimal, flexible, and automated.

Multi-agent-based modeling techniques have been used to integrate real-time intelligent decision-making in building control. For instance, an indoor environment that actively supports its inhabitants can be created with these techniques [57]. These modeling techniques also include unpredictable user-behavior, fluctuating weather conditions, and grid imbalances [52,58]. For instance, the study by Anvari-Moghaddam et al. [52] demonstrates how MAS is used to optimize management strategies for a building through computer simulations in combination with third-party software such as MATLAB and GAMS. Hence, studies show that energy consumption can be reduced without compromising the inhabitants' comfort level in residential buildings.

In the study [52], a smart grid is simulated with several residential buildings, conventional and RES. The residential buildings include underfloor heating, heat pumps, and energy storages. The simulation incorporates meteorological data for the examined location together with technical data, to estimate the power production from RES. The simulation result shows that it is possible to reduce domestic energy consumption and meet the system's objectives and constraints at the same time. However, the study does not take fault-tolerant and uncertainty handling capabilities into account.

The study by Zeiler and Boxem [16] analyses how smart grid and building optimization can work together and presents an ontology of a software system which acts as a bridge between BMS and a smart grid. Several experiments are conducted in this study to test a HVAC system in a building environment, including the interaction with a smart grid. The study also includes the dynamic behavior of the occupants towards the systems in combination with an overall goal of energy efficiency. The study finds that different elements depend on each other, e.g., changes in required heating affect the available energy. The automated equipment, controlled and managed by the building, responds to demand response requests from the grid to balance the grid condition [59]. The experiment also shows that the comfort level increases while the energy consumption decreases in their MAS modeling.

Meanwhile, the study by Mousavi et al. [53] includes the unpredictable nature of the business process in an office building in a simple model with only a few devices to control. This study does not include a response to the grid conditions. Instead, the study investigates an energy automatic model for office buildings to reduce energy consumption and increase the indoor comfort level. The model is a MAS with the ontology based on the standard IEC 61499 (automation system standard) [60]. The goal of this study is to optimize the energy consumption in an office building where the ontology provides the communication logic and allows agents in the model to share knowledge and data [61]. In the MAS model, agents communicate and collaborate towards a common goal. The method has been applied to an office meeting room, where meeting activities and equipment can be automatically controlled, including measurements of energy consumption. Based on the data gathered as a result of the simulation, the study shows that it is possible to reduce 50 % of the room's monthly energy consumption by controlling the operation and preparation of the room automatically. The duration of the meeting room simulation is 20 working days (1 working month). The simulated BMS automatically acknowledges the meeting schedules and needs for shading, screen, and blackboard usage, etc. The business process is combined with automated processes to overcome the inefficient use of energy in buildings and lower the number of system failures.

3.2.4. MAS Tools for the Energy Domain

In a MAS of the energy system, agents can represent market players, network components, or part of/a whole system [9]. Therefore, the multi-agent architecture of energy and power systems is designed for dealing with the system complexity [9,23]. Meanwhile, multi-agent simulations allow investigating the statics and changes of the physical systems, electricity market and market players' behaviors. There are multi-agent simulators in the various domain for different purposes, e.g., CoABS (https://www.cs.cmu.edu/~{}softagents/project_grants_coabs.html) grid [62]. The selected literature shows that the multi-agent simulators in the energy system can be divided into three main areas:

1. Multi-agent simulators for smart grid:

    - Mosaik (https://mosaik.offis.de/): [49,50] is a flexible smart grid co-simulation framework, and allows to reuse and combine existing simulation models and simulators to create large-scale smart grid scenarios [63]
    - MASGriP (Multi-Agents Smart Grid Simulation Platform): models the internal operation of a smart grid with the consideration of all involved players [21].

2. Multi-agent simulators for the grid communication, monitoring, and control:

- Electric Power and Communication Synchronizing Simulator (EPOCHS) (http://www.cs. cornell.edu/hopkik/epochs.htm): aims to solve network communication problems and avoid potential costs and damages by the combination of the results of several simulators [64].
- Global Event-Driven Co-Simulation framework (GECO): models and simulates the control, monitoring, and protection of the power systems and communication network [65].

3. Multi-agent simulators for electricity markets:

- Multi-Agent Simulator for Electricity Markets (MASCEM) (http://www.mascem.gecad.isep. ipp.pt/overview.php/): can simulate many market models and player types, and enable decision-support [21].
- Agent-based Modeling of Electricity Systems (AMES) (http://www2.econ.iastate.edu/tesfatsi/ AMESMarketHome.htm): simulates wholesale power market operation including load, market participants, grid [66].
- Power Trading Agent Competition (Power TAC) (https://powertac.org/): is an open-source platform that simulates future electricity market including broker types of energy retailers, commercial or municipal utilities, or cooperatives [67].
- Electricity Market Complex Adaptive System (EMCAS) (https://ceeesa.es.anl.gov/projects/ emcas.html): simulates diverse participants' strategies and behaviors in the electricity market [68]
- Multi-Agent Negotiation and Risk Management in Electricity Markets (MAN-REM): simulates electricity markets, and emphases the bilateral contracting and risk management [37].
- Adaptive Learning strategic Bidding System (ALBidS): aims to integrate market strategies, evaluate performances under different contexts of negotiation, and provides decision support to electricity markets negotiating players [69].

3.3. Ontology and Defined Ontologies in the Energy Domain

3.3.1. Definition of Ontology

The term 'ontology' is originally introduced by the Greek philosopher Aristotle [70] as a theory about the nature of existence. Since the beginning of the 1990s, ontology has been adopted by information scientists in the field of artificial intelligence and web and system modeling [71]. In computer science, the ontology is defined as: "a formal, explicit specification of a shared conceptualization." [72]. This explicit formal specification is domain-specific [73]. Ontology provides a model to support the process in agreement with all parties that all parties commonly agree to refer to the 'specification' of a conceptualization [74]. Uschold [75] identified different categories of ontologies:

- Communication between people. Here, an unambiguous but informal ontology may be sufficient.
- Inter-operability among systems achieved by translating between different modeling methods, paradigms, languages and software tools;

In the Artificial Intelligence community, ontologies describe entities and their properties, relationships, constraints and behavior that are not only machine-readable but also machine-understandable [14,24]. According to [13], the functions of ontology are:

- Communication: ontology can provide common glossaries to communication among different individuals.
- Interoperation: ontology can freely interpret and map among various modeling methods, languages and software tools.
- Reuse: the ontology's analyses clarify the structure of the field's knowledge in order to lay a good foundation for knowledge representation. Ontology can be reused, so the repetitious knowledge analyses can be avoided.

- Knowledge acquisition and sharing: to construct the system based on knowledge, the available ontology can be used as origination and foundation to supervise the acquisition of knowledge, which can improve its velocity and reliability.

To build an ontology, knowledge engineers need to talk with domain experts to analyze the system and to make everything explicit, e.g., concept description with existing defined concepts and the knowledge rules (i.e., the decision-making rules) in these formalized concepts [76]. There are seven recommended steps to design an appropriate ontology. The developed ontologies provide the means to exchange information that can be interpreted by software agents, knowledge representation and sharing among the software agents [38]. Ontologies are also useful for sharing between modelers, domain, experts, and users [14]. Meanwhile, ontologies also enable to infer knowledge from the gathered information using a reasoner [38].

Many languages have been developed to build an ontology for different purposes. The Ontology Web Language (OWL)(https://www.w3.org/OWL/) by W3C is one of the most popular standard ontology languages. It possible to use OWL in a variety of applications such as knowledge sharing and representation [77], semantic web [78], information system [79], ontology-based reasoning [80], etc. An important requirement for the system interoperability is to reuse existing ontologies. There are some libraries of reusable ontologies available online, such as Ontolingua (http://www.ksl.stanford.edu/software/ontolingua/) and DAML ontology libraries (http://www.daml.org/ontologies/) [48].

3.3.2. Defined Ontologies in the Energy Domain

There are some ontologies already developed for specific energy domains [81,82]. For instance, Kofler et al developed an ontology that focuses on energy consumption and energy provision [83]. Ma et al. [84] proposes 6 basal ontologies for energy management system:

- Cognitive ontology: the activity that agents analyze power systems.
- Physical entity ontology: the equipment that is used for transmitting electric energy and its connecting topology.
- Data ontology: the magnitude that cognitive agent has apperceived to respond to physical entities.
- State ontology: the generalization of the current operation mode in an electric power grid.
- Event ontology: all aspects that create changes of state.
- Operation Ontology: the combination of all actual actions that a cognitive agent does on physical entities.

The well-described ontologies in the energy domain are mainly found in the electricity market domain. For instance, [85] develops an ontology for the electricity market named Electricity Market Ontology (ELMO). It provides a shared, common understanding of concepts and procedures in the electricity market operation. The ELMO ontology uses a multi-layered architecture divided into highly maintainable, extendible, and reusable modules that can be used by organizations such as the Hellenic Transmission System Operator (HTSO). The ontology is primarily developed specifically for the electricity market of Greece, and the adaptation to other markets are thereby difficult.

Other examples are the studies of Santos et al. [3,47,48] that develop an Electricity Market Ontology (EMO). The EMO is an upper ontology for the electricity market, from which other low-level ontologies can be extended. It defines the main concepts of the electricity market, and the specific ontologies extended from the EMO define requests, responses, and notifications. Ontologies for the EPEX [3] and Nord Pool spot market [47] are developed as extensions of EMO. The research in [48] states that the aims of EMO are to be extendable and reusable in the development of other low-level ontologies for specific markets, such as MIBEL or IPEX (The Belgian and Dutch electricity market).

3.3.3. Ontology Design

Gruber provides five design principles [86] for the development of ontologies: clarity, coherence, extendibility, minimal encoding bias, and minimal ontological commitment. For ontology design, it is necessary to consider the ontology representation languages including tools to create and manage ontologies. Some standard ontology languages have been established with stable tools for the Semantic Web community, e.g., the Resource Description Framework (RDF) [87], RDF schema (RDFS) [88] and the Web Ontology Language (OWL) [88].

Several features of the Semantic Web languages are important for the ontology development, e.g., Open World Assumption (OWA), Description Logics (DL), and service representation. OWA assumes that knowledge is always incomplete. It is very important because incomplete information is common, and fragments of knowledge are often distributed within multiple ontologies [89]. Comparatively, the Closed World Assumption (CWA) assumes that if a statement cannot be proved to be true then it is false. DLs are formal languages designed for knowledge description and standard reasoning and provide the underlying formal framework for OWL and RDF [90]. DLs are known as the basis for ontology languages and are used to define, integrate, and maintain ontologies [89]. DLs are discussed in [24,48,91].

Together with the introduction of the ontology design. Semantic web services are an integral part of the Semantic Web and aim to be automatically discovered and invoked by computer programs [92]. Therefore, semantic web services must be able to describe the provided information and how this information can be retrieved [93]. A number of languages are available to describe services, e.g., OWL-S [94], Web Service Modeling Ontology (WSMO) [95], WSDL-S [96], and FLOW [97].

- Categories of ontologies

Ontologies can be categorized into three levels: upper ontologies, domain ontologies, and application ontology (shown in Figure 2) [98]. Upper ontologies provide common and consistent concepts that are referenced by other ontologies. Several upper ontologies exist, e.g., Suggested Upper Merged Ontology (SUMO) [99] and DOLCE [98]. Domain ontologies reuse or specialize concepts from the upper ontologies, and specify terms, relationships that are relevant in a particular domain. For instance, the domain ontology in [15] describes the concepts of the process dynamics, control, automation and the services provided by the agents, and defines relevant classes of entities and relations between entities. Application ontologies re-use and extend terms from one or more domain ontologies to apply for a specific application, and generally cannot be reused for other applications.

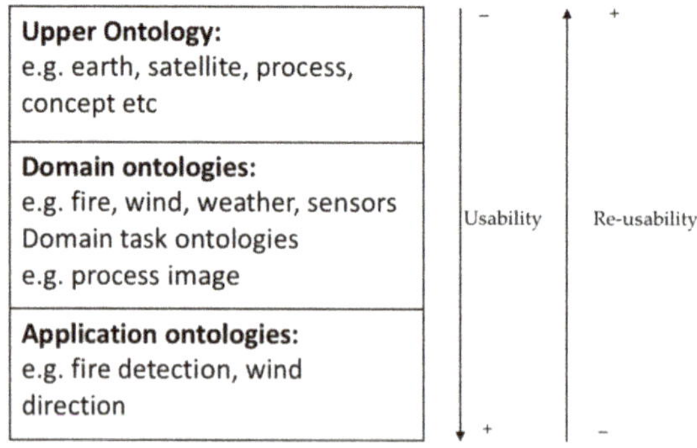

Figure 2. Three ontology levels [98].

In the energy domain, ontologies for complex systems are often separated into a hierarchy consisting of an upper ontology that is connected to several lower-level ontologies representing specific subdomains [100]. The three MASs (MASCEM, ALBidS, and MASGriP) developed by

Santos et al. [21,101,102]. Are all framed by an upper ontology, which allows communication between the simulations. However, this approach requires universal acceptance from all entities involved, and the low-level ontology for each layer still needs to be extended. Dam, Nikolic, and Lukszo [103] propose the generic ontology and the case-specific ontology, where the case-specific ontology is a specialization of the generic one and the generic ontology is a generalization of all underlying case-specific classes shown in Figure 3. Dam, Nikolic, and Lukszo [103] also suggest how to decide on the borders of the generic and domain-specific class in ontology. In [85], the ontology is divided into smaller building blocks, which makes it easier to modify and reuse in other models. [16] proposes a hierarchical ontology for the energy supply structure of buildings (shown in Figure 4). This proposed hierarchical ontology aims to investigate the interaction between energy flows on different aggregation levels within a building.

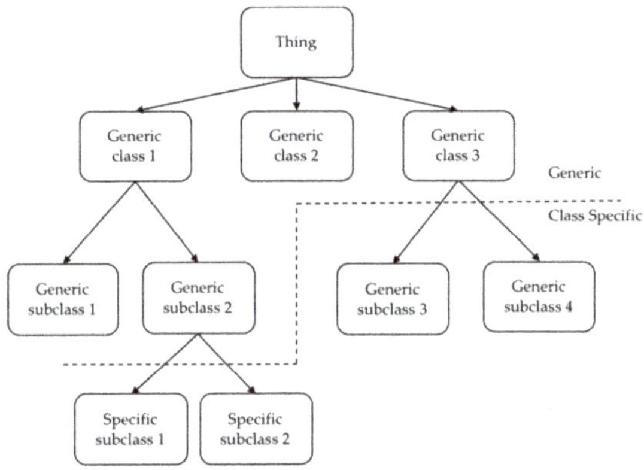

Figure 3. The border between generic and domain-specific class in an ontology [103].

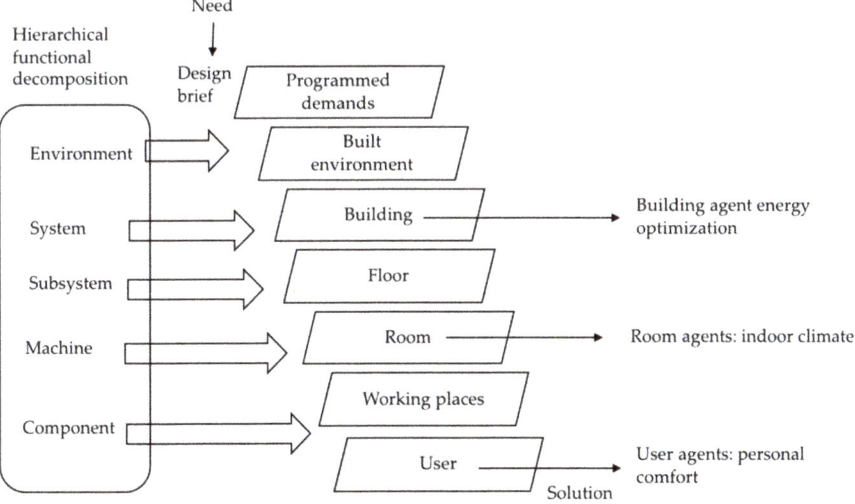

Figure 4. A hierarchical ontology for the energy supply structure of buildings [16].

- Ontology mapping

MASs are usually developed independently and may not use the same upper ontologies. Therefore, ontology mapping is needed when an application requires access to multiple individually created ontologies. The mapping between ontologies can take much work [104], and there are several ontology mapping tools available [105]. Among the selected literature, ontology mapping is discussed and presented in [9,21,49,91].

- Ontology development tools

There are many tools for developing ontologies [106], e.g., Protégé (https://protege.stanford.edu/) and SWOOP [107]. Protege is well established and used by a large user community. For instance, Protege is used in the selected literature [23,24,48,50].

- Ontology development process

Ontology development processes is a relatively new field of study, including ontology life cycles, methods, and methodologies for building ontologies [89]. [108] introduces a methodology for ontology development including three phases: specification, conceptualization, and implementation. Noy and McGuinness [76] propose a more detail and practical ontology development process with seven steps which have been popularly used:

Step 1. Determine the domain and scope of the ontology
Step 2. Consider reusing existing ontologies
Step 3. Enumerate important terms in the ontology
Step 4. Define the classes and the class hierarchy
Step 5. Define the properties of classes—slots
Step 6. Define the facets of the slots
Step 7. Create instances

3.4. MAS Design and Architectures

3.4.1. MAS Design Methodologies

According to [109], the MAS design usually consists of

(1)	A conceptualization phase where the problem to be solved is specified;
(2)	An analysis phase;
(3)	A design phase that uses the results of the analysis phase to produce agent designs of varying detail

Although the majority of the selected literature not specifically present their phases of the MAS design methodology, the introduction of the MAS architecture/structure in their cases is more or less according to the Gaia methodology (shown in Figure 5). The Gaia methodology is popularly adopted for the analysis and design of the agent-based system, it is used in [23,24,50]. Some other similar methodologies are also used for the agent-based system design, e.g., High-Level and Intermediate Models for Agent-oriented Methodology (HLIM), Modelling Agents and their environment (AUML), MASE [24].

Another MAS design methodology proposed by the IEEE PES MAS working group(http://sites. ieee.org/pes-mas/agent-technology/design/) is mentioned in [17]. This MAS design methodology is proposed by [110] with six stages, and each stage of the methodology produces material that is input to the next stage (shown in Figure 6):

- Requirements and knowledge capture stage: the MAS design usually begins with a particular problem. To solve this problem, this stage specifies the system requirements and capture the knowledge needed to fulfill those requirements. The system requirements and captured knowledge is the input to the next stage.

- Task decomposition stage: it transforms the requirements specification and captured knowledge from the previous stage into a hierarchy of tasks and subtasks. These tasks may include the functions performed by legacy systems.
- Ontology design
- Agent modeling stage: based on the task hierarchy and ontology design, it identifies a group of autonomous agents performing the tasks in the task hierarchy. Each task in the hierarchy must be attributed to at least one agent and one agent can encapsulate one or more tasks. The outcome is a set of agent models that specify the tasks the agents perform. The tasks attributed to legacy systems and generated new codes are also identified at this stage.
- Agent interaction modeling stage: it defines the interactions the identified agents support. The output usually is the interaction diagrams.
- Specification of agent behaviors stage: it specifies the interaction functionality of the agent and the control functionality of the agent.

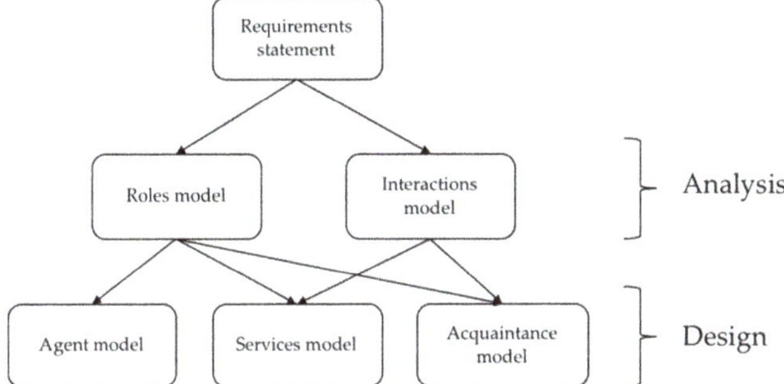

Figure 5. The conceptual framework of the Gaia methodology [111].

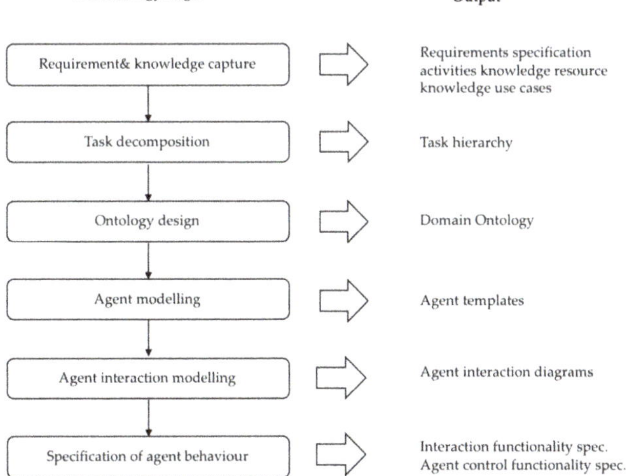

Figure 6. Agent design methodology stages and their output used during the design of the PEDA (Protection Engineering Diagnostic Agents) system [110].

Some MAS design also defines layers of the MAS architecture, e.g., [23] that the MAS system architecture includes two layers: the management layer and the subjacent execution layer. The management layer is responsible for functionalities that can be considered general in the frame and are covered by the agent control systems. The subjacent execution layer employs the automation agents' tasks.

- MAS Development environment

The MAS development environment in the selected literature is usually performed using JADE (Java Agent Development Environment) [21], e.g., in [17,23,24]. JADE is one of the agent platforms compliant with FIPA standards. JADE provides services such as agent management system, directory facilitator, agent communication channel, etc., and supports the paradigm of avoiding SPOFs (Single Point of Failures). JADE has limited support for Semantic Web technologies. Therefore, some extensions are usually used to compensate for this issue. For instance, AgentOWL provides support for OWL ontologies using JADE agents [112] and AgentScape attempts to deal with scalability issues [113].

3.4.2. MAS Architectures

The MAS architectures (sometimes also called MAS structure, or MAS organizational structure) in the selected literature includes agent types and agent management framework/ system architecture, followed by the agent communication and ontology design that are usually introduced together, sometimes with agent interaction/activity diagrams.

- Agent types

The agent types are defined based on the system requirements, captured knowledge, and decomposed tasks. For example, agent types represent the devices and units in a power system, e.g., building management agent and RES (Renewable Energy Resources) agent [8], distributed voting agent and monitoring/resurrection agent [20], bus agent and switch agent [23]. In some MAS, the agents control the corresponding equipment according to their objectives, the measured and collected data, etc.

Agent types represent market players that especially for the electricity market, e.g., user agents and energy market control agent [17], system operator agent and VPP (Virtual Power Player) agents [37]. Some agent types are also ontology related agents, e.g., translator agent and ontology Agent [53]. In [53], the translator agent communicates with the main controller function blocks, and the ontology agent extracts knowledge from the ontology-based on requests.

Sometimes, the agent goals are also introduced together with the agent types. For instance, in [8], the agents act to achieve three goals of system load supplying, energy cost minimization, and residents' comfort maintenance.

- Agent management framework/ system architecture

The structure of a MAS usually is illustrated in the agent management framework/ system architecture. For instance, the proposed agent management framework [8] including several components, e.g., the agent platform, agent container, and directory facilitator, etc. The system architecture also can visualize the multi-layered structure, e.g., [20], and the relations of agents and environment, e.g., [17], and relations of agents and physical systems, e.g., [23].

3.5. The Application of Ontology in MAS Development

3.5.1. MAS Interoperability and Ontology

In a MAS, it is important to set up a communication language for meaningful conversations between agents. The agents communicate through message exchange so-called Agent Communication Language (ACL). ACL is the existing interaction language standard for exchanging knowledge between agents. For a given Communication act «F(P)», the «F» part refers to the MAS and is regulated by the ACL standard, and the «P» refers to the domain knowledge. In our case, «P» refers to the «Energy Domain» or the «Energy Ontology» [114].

However, even an agent development environment supporting the same Agent Communication Language (ACL) and content language are implemented in two MASs, it does not mean that the agents in the two MASs can share any useful information because different ontologies are used in the two MASs [91].

As more applications of MAS in the energy domain for advanced functions and MASs are not expected to operate in isolation from each other, the interoperability challenge raises due to an increasing requirement for data and information exchange between systems. Therefore, there is a need for full interoperability and open standards for the MASs in the energy domain [91]. The interoperation issues of existing multi-agent systems have highlighted in the literature, particularly the issues of the use of different ontologies. Meanwhile, it is important to establish the same language, especially a common ontology for the communication between agents.

3.5.2. Agent Communication and Ontology

Agent communication in MAS can be accomplished in two ways: immediate communication among agents and interaction in a unitive environment [13]. MAS usually implements higher-level communication and supports reasoning abilities based on the Agent Communication Language (ACL) and a common vocabulary defined in an ontology [115]. The agent communication and ontology design in the selected literature is similar to the combined stages of ontology design, agent modeling, agent interaction, and specification of agent behaviors stages proposed in [110], and usually consist of standards for agent communication, interoperability, and ontology design.

- Standards for agent communication and interoperability

A standard for the communication between agents has been proposed by the Foundation for the Intelligent Physical Agent (FIPA [116]. The FIPA standards have been popularly used by MAS developers in the computer science community and FIPA was formally accepted as a standards committee of the IEEE Computer Society In 2005 [109]. Such standardization promotes open specifications for the interoperability between agents and MAS [117]. The FIPA standards include specifications for the agent communication language, communicative acts, content languages, and message transport protocols. It also includes a standard that proscribes the agents that a MAS must implement to be FIPA compliant

The FIPA-ACL specifies the syntax, the content of the message provides the semantics of the message including the content language and the ontology [118]. The messages built under the ACL structure allow the definition of various elements (e.g., performative, sender, receiver, content, language, and ontology, among others) and various communicative acts (e.g., agree, cancel, confirm, not-understood, etc.) [9]. Meanwhile, the correct interpretation of the meaning of the message is assured, the ambiguity is removed about the content [21]. The MASs in [3,47,48,52] all apply the FIPA-ACL. There are other ACL investigated in the literature as well [119,120], e.g., Open Agent Architecture (OAA) in the work of Praca et al. [42].

MASs developed by different platforms can interoperate with these FIPA standards, but it doesn't mean that useful information can be shared between agents if the MASs employ different ontologies [21,91]. It requires MASs share a common vocabulary, so the messages may be interpreted correctly among agents [47]. Therefore, ontologies are used to enabling the standardization of communications and interpretation of concepts between MASs [48].

The IEEE standards committee has identified the challenge of interoperable protocols, data formats and meaning and stated that open communication between smart devices using common protocols is crucial to interoperability [121]. Some standards in the power systems promote interoperability between devices within substations and open interfaces between energy management systems [91,109]. The most widely applied standard in the power system is the IEC 61970 Common Information Model (CIM), and its distribution management extension IEC 61968 [122].

IEC 61970 Standard is proposed by the International Electrotechnical Commission (IEC) to discuss and plan a variety of electrician and electron standards in order to procure international cooperation. IEC 61970 Standard defines the application program interface (API) of the energy management system is promulgated by IEC No.57 technical commission (Group 13) [13]. There are five main parts in the IEC 61970 standard: introduction and basic request, glossary, common information model (CIM), and two levels of component interface specification (CIS).

The CIM is a three-layer domain model, it defines a common vocabulary to describe the basic components used in electricity transportation and distribution [38], and CIM aims to facilitate power management processes (e.g., outage management, asset management, and customer information management) [50].

To achieve coherent and advantageous cooperation between different power systems, some reference models and frameworks are also popular used, e.g., SGAM (https://sgam-toolbox.org/) (the Smart Grid Architectural Model), USEF (https://www.usef.energy/) (the Universal Smart Energy Framework), and SEAS knowledge model (https://www.the-smart-energy.com/) (Smart Energy Aware Systems).

The Open Automated Demand Response (OpenADR) (https://www.openadr.org/) and energy@home (http://www.energy-home.it/SitePages/Home.aspx) models are also highly discussed in the literature. However, [50] states that 'none of these standards cover the whole semantics involved in a flexible urban energy network on its own, and they are not formally aligned with each other'. For example, the term 'equipment' could refer to transmission system equipment, or domestic appliance equipment [50].

- Ontology-based agent communication design

According to FIPA [107], semantic MAS interaction can be specified with three dimensions: 1) Internal agent behavior: action selection and execution; 2) External (agent) interaction to exchange: a) content of the interaction including both information and tasks; b) context of the Interaction and its relation to an agent organization; 3) System, or platform, services: message transport, discovery, action execution, management, and inter-platform interaction.

The FIPA (agent interaction) model (often referred to as the FIPA-ACL) is an Agent Interaction Protocol Suite (AIPS). The AIPS contains several distinct semantic protocols for agent communication including interaction process, communicative acts, content logic, and content ontologies (shown in Figure 7) [107].

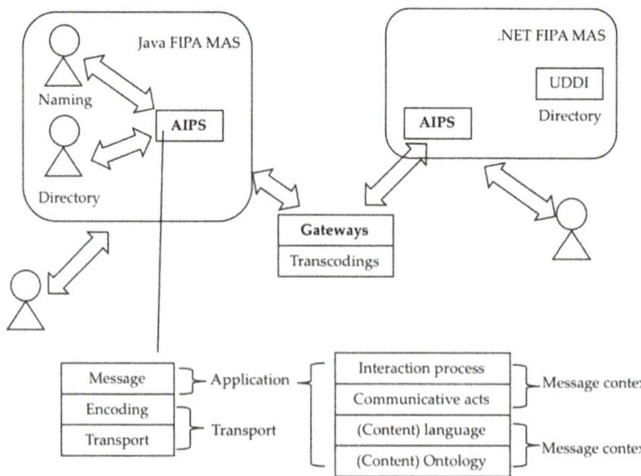

Figure 7. Foundation for the Intelligent Physical Agent (FIPA) specifies multi-agent systems (MAS) interaction using specifications for an Agent Interaction Protocol Suite (AIPS) and MAS platform [107].

The design of internal agent behavior and interaction in a MAS mainly concerns the agent communication models as in the majority of the selected literature. The design of agent communication usually includes messages (message content) and message exchange (protocol). Messages and protocol are usually described in the UML diagrams as class diagrams and sequence diagrams, e.g., in [20], the communication sequence and communication parameters are introduced. The content of a message comprises two parts: content language (provides the syntax or grammar of the content) and ontology (consists of the semantics or lexicon of a message) [91]. The ontology-based agent communication model can be shown in Figure 8.

MAS developers usually use JADE to create agents because JADE agents communicate by exchanging message in compliance with the FIPA ACL. The FIPA Semantic Language (FIPA-SL) is popularly adopted as the standard content language [123]. In FIPA-SL, an ontology comprises a list of concepts, predicates, and actions specific to the domain of communication. However, the structures of ontologies in the selected literature are different. For instance, the ontology in [124] is defined in the form of EBNF and includes seven parts (policy, modality, trigger, subject, behavior, target, and constraint), and the ontology in [20] contains four parts (ID, type, parameter, and value).

When designing a MAS, developers usually introduce the syntax and semantics of the domain ontologies and application-specific ontologies applied in the MAS and describe the purposes and functions of the ontologies. For instance, [124] applies a policy ontology in their MAS. In [124], the policy ontology regulates behaviors of agents including application activity, authorization activity, monitoring activity, requesting-monitoring activity, discovery activity, and negotiation activity. This research designs a policy engine within each agent who is the subject of obligation policies or the target of authorization policies and the policy engine interprets and enforces the policy when the policy is enabled.

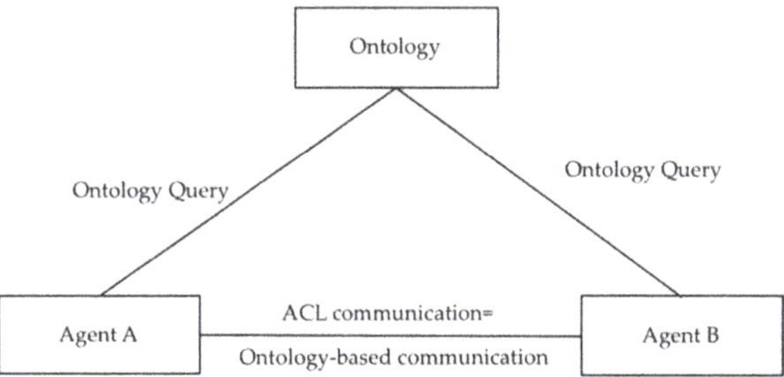

Figure 8. The illustration of the ontology-based agent communication model [125].

The FIPA agent standards focus on specifying protocols for external interaction and platform services rather than on the internal agent behavior [111]. It is because the internal agent behavior is are often problem-specific or application specific, and not easily accessible and observable. In the FIPA Ontology Service, an ontology agent is recommended to provide a number of ontology-related services for solving the problem of using multiple ontologies [91].

However, this solution is difficult to be implemented due to challenges of the system integration including between-ontology mapping, translation mappings, etc. Therefore, [91] recommends defining a common upper ontology that represents the general concepts used in the domain of power system. Meanwhile, related common standards in a domain can serve as a foundation for an upper ontology, e.g., The power systems Common Information Model (CIM) [126]. The upper ontology for the MAS interoperability of the electricity markets and demand side is well discussed by Santos et al. [3,47,48].

4. Discussion

The literature shows that there is an increase in MAS application in energy domain since the distributed nature of MAS allows the energy system design to deal with complex systems [127].

In a MAS, complete knowledge about the system is not required, but each agent in the system acts autonomously toward some predefined objectives to optimize the system performance [128]. Therefore, agents have possibilities to represent different market participants, network components, or systems [9]. The agents' individual goals decide the agents' behaviors, e.g., either cooperate or compete with other agents [127]. The behavior of the overall system is a result of the agents' behaviors.

MAS is not necessarily a simulation tool, but simulations may be important for the study of the energy domain, e.g., scenario comparisons, evolution studies, and sensitivity analyses. Several MAS studies are found in the literature dedicated to the energy domain. For instance, the study of Koritarov [1] demonstrates the application of the EMCAS in electricity markets. The model enables the investigation of the physical infrastructure and the economic behaviors of the market participants. The study of Li et al. [39] demonstrates the AMES simulation for the wholesale operations and market participates strategies. In the building sector, the simulations of electricity consumption in an office building are simulated by Mousavi et al. [53]. The study considers the unpredictable nature of business processes. Meanwhile, the research by Zeiler and Boxem [16] simulate the grid conditions in their study of building control.

All these simulations aim to solve problems in specific domains and are limited to an existing system (do not allow for connections to external systems) or do not take advantage of the formal exchange of knowledge. It is possible to solve problems that cover more complex domains if these systems can communicate and exchange knowledge with each other.

The combination of different systems can simulate a complex system such as the energy system. In such a system, stakeholders work together, interact, and negotiate with each other, while the demand and supply of resources need to be managed. The heterogeneity among these systems makes the interoperability complex, and the system may have different domains, concepts definitions, programming languages, etc. In order for the MAS to be able to communicate with each other and overcome their individual limitations, a mechanism for communication is important. This mechanism should allow information and knowledge sharing. At the same time, the system should be flexible to deal with several processes. Therefore, a communication standard should be defined, ensuring that agents in the system use terms with the same meanings [129].

The FIPA is the de facto standard for agent development [9]. FIPA provides different interoperability standards, e.g., the standard agent communication language (FIPA-ACL), which make it possible to integrate different MASs [130]. However, it does not mean that agents belonging to different MASs can share any useful information if the MASs use different ontologies. The ACL provides a framework for the communication standardization between agents, but the standard only defines the structure of messages and interactions. Therefore, agents speak the same language but do not share the same vocabulary.

In an ACL, the content of messages must be understood by agents for the messages to be meaningful. Catterson et al. [91] describe it as " … the structure and meaning of the content are in a format expected by the receiving agent so it can decode the sender's intentions". Agents exchange information to achieve their goals and therefore must apply the same language to interact with each other. But it also needs a common representation of concepts for agents, which ontology can provide.

The ontology describes the concepts and the relations among agents and therefore must be a part of each agents' knowledge base [131]. Ontology is described as a form of knowledge representation of the world or some parts of it and "provides a shared vocabulary, which can be used to model a domain that is, the type of objects, and/or concepts that exist, and their properties and relations" [132]. Meanwhile, Luncean et al. [131] states that "An ontology is used to represent knowledge that is shared between different entities. It provides terms and vocabulary used to represent knowledge so that both sender and receiver can understand" Several ontologies already exist in the energy field. In [16,52], the main goal of ontologies is to support the interactions between energy management of buildings and the smart grid.

It is important to mention that the design of an ontology itself does not contribute to energy savings or energy-neutral building environments. However, it brings several benefits to the design of the software process of a MAS. First, it gives a deeper insight into the modeled domain and system functionality. Secondly, it reflects upon the data types and required communication between agents. These factors are useful when concepts are shared between different teams and systems, e.g., when different domains need to be connected to a smart grid.

However, MASs in the energy domain are developed with their own ontology, which cannot be directly integrated into other systems. A standard to solve the problem of multiple ontologies would lower the cost and human effort when different systems need to be connected. In the literature [91], several solutions for MAS integration are investigated. The FIPA ontology services the integration of existing MASs by introducing an ontology agent. This agent provides ontology-related services, e.g., translating expressions between ontologies and identifying a common ontology to two agents [16]. However, ontology designers still need to identify the similarities and differences between ontologies manually to translate the ontologies. This likely introduces more complexity and potential errors.

An upper ontology, as discussed in [21,103] could be an alternative to represent the general concepts of the domain. Such ontologies provide the framework in which the low-level ontologies can work. The upper ontology allows communication between different systems and each system with separated low-level ontologies. An upper ontology can be defined through multi-layered architecture or smaller reusable modules. The development and maintenance of MAS are easier and more efficient by composing a large-scale ontology out of smaller ones. This makes the ontologies simpler to modify, e.g., if legislation changes. The independent parts of an ontology must be well defined and separated. Thus, it is possible to reuse the parts in similar applications. The layered architecture also makes the ontology easier to be extended for other application domains and not just the intended domain [91].

An upper ontology for the energy sector can serve as an open standard that can assist the development of multi-agent solutions. It should not be a standard for all applications, but a tool from which the low-level ontologies can be extracted. Upper ontologies for the electricity domain are found in the literature, but the integration with the entire energy sector is still missing. This integration is necessary to fully understand and control the energy sector because the energy sector becomes more complex and consists of multiple hybrid systems.

The literature reviewed in this study presents different energy domains and includes different agents, data, and terms. This heterogeneity hinders the full adoption of these MASs and ontologies in a real scenario. Hence, there is a need for developing a unified ontology that represents all energy domains and provides a common terminology. In the literature, business models are separated from the MASs in the energy domain. For a deeper understanding of the domain and related agents, business models should be considered as part of MASs.

The combination of MASs, ontologies, and business models will enable simulations of the energy sector for exploring the interplay of policy, economy, and technology. Furthermore, a standardization of communication between agent will provide better knowledge- and data exchange between agent and domains. However, better simulation tools which can be used for scenario comparison, prediction of future evolution and sensitivity analysis are important, and it will make simulations easier to predict future events, identify unmet needs and act deliberated to changes in the energy sector.

5. Conclusions

This study contributes a scoping review of literature on the application of ontology in the MAS for the energy domain. It is evident from the literature highlighted in this study that multi-agent ontology approaches are of emerging interests in the energy sector and that complex system modeling is an essential tool in assessing control strategies and new policies for designing more efficient systems.

The selected publications show that the application of ontologies in the field of MAS for the energy domain was mainly conducted after the year 2004, focuses on the sub-domain of grid control between

2004 to 2014, and mushrooms into the sub-domain of electricity market since 2014. The discussion of ontology and MAS in the selected publications can be divided into five categories:

- Definition of agent and MAS. The definitions of agent, intelligent agent and MAS and the introduction of an agent structure are given in some selected publication. However, some publications do not differentiate the agent-based system and multi-agent-based systems.
- MAS applied energy domains. The applied energy domains include grid control (also, microgrids), electricity markets, demand-side and building systems. The applied MAS tools are also introduced in some selected publication.
- Defined ontologies in the energy domain. Definition of ontology, functions of ontology and the defined ontologies in the energy domain are introduced. The ontology design is introduced usually together with the agent communication model. Although generic ontology and the case-specific ontology, upper-level, and lower-level ontology, and ontology hierarchical are introduced, a systematic discussion on the categories of ontologies (upper ontologies, domain ontologies, and application ontology) is missing. Meanwhile, although ontology mapping for inter-MAS communication and ontology development tools are introduced, the ontology development process is not yet discussed in the selected literature.
- MAS Design and architectures. The MAS design methodology-Gaia methodology is introduced and applied in some selected publication, and MAS design methodology proposed by the IEEE PES MAS working group is introduced but not well discussed or applied in the selected publication. The MAS Development environment, JADE, and its extensions are introduced but the design detail with JADE is missing. The MAS architecture is commonly introduced with the description of agent types and agent management framework/ system architecture.
- Ontology in the MAS development. The importance of ontology for the MAS interoperability is emphasized and the application of ontology in the agent communication design is well discussed in the majority of the selected publication. The standards for agent communication and interoperability are discussed with two dimensions: standards for domain-specific, e.g., the SGAM reference model, the power systems CIM and SEAS knowledge model in the energy domain are discussed; The FIPA-ACL is applied for almost all MAS design in the selected publication.

5.1. Recommendation of the Ontology-Driven MAS Development for the Energy Domain

Based on the review result, this paper finds out the following aspects in the ontology-driven MAS development for the energy domain should be further discussed, developed or emphasized:

- The ontology development process in MAS design

Although the importance of ontology in the energy domain has been emphasized, especially for the MAS interoperability. However, from the ontology engineering perspective, the ontology development process has not been addressed well in the MAS design, especially with the consideration of the ontology categories. This paper recommends the further work can combine the categories of ontology [98] and the ontology development process [76] into the MAS design with two aspects: multi-agent communication and MAS interoperability.

- The detail design process and realization of the ontology-driven MAS development

The selected publications well discuss the 'what' and 'why' of their designed/developed MASs. However, the 'how' is missing in the majority of the selected publication. Therefore, it is difficult for readers to re-produce their methodologies of MAS development. Therefore, this paper recommends the further work can focus on this aspect, and it is especially important for the MAS interoperability.

- Open standard implementation and adoption

Open standards for both MAS design, agent communication and energy domain are discussed in the selected publication, especially regarding the MAS interoperability. This paper finds that the MAS interoperability issue is not solely due to the inter-MAS communication barriers, and upper ontology design cannot solve this issue if the designed upper ontology or the selected open standard is not adopted by other MASs. Therefore, further work on the open standard implementation and adoption for the ontology-driven MAS development is recommended.

- Higher intelligent MAS development

The MAS interoperability is important for the distributed energy systems, and ontology improvement (upper ontology or generic ontology) seems like the only solution in the majority of the selected publications. This paper recommends the future work can consider developing higher intelligent MASs that allow the 'fuzzy communication' between MASs.

- Inter-domain MAS development

Although this paper tries to search literature in the energy domain for both electricity and heating. However, the search result only shows in the electricity domain, and the literature on MAS and ontologies for the heating sector is missing. Heating is an important subdomain in the energy sector and is also strongly connected to the electricity sector through combined heat and power generation, and electrical heating. Hence, heating should be equally addressed in the studies of MAS and ontologies for the entire energy sector. The priority for future work in this field should focus on the interoperability with further external systems and cover the simulation of other areas in the energy system, including heating. However, the inter-domain ontology design will be more complex and difficult compared to only under the electricity-related domain.

- Agent listing

Agent types, roles, and interactions are well introduced in the selected publication. Meanwhile, the domain analysis in the MAS design methodologies is introduced. Some studies have done illustrations of agents in smaller scales, e.g., [3,47,48,52]. However, a systematic approach to list all related agents with a clear MAS boundary is missing. In a MAS, agents are specialized to perform tasks based on their individual goals [133]. Meanwhile, a MAS with stakeholder listing can give a good overview of the whole system. The literature shows that there are different ways to illustrate the identified agents together with their relationships. Some authors [53,81] introduce agents with descriptions, and others [18,52] use diagrams to graphically present agents. One example of the graphical illustration is the Harmonised Electricity Market Role Model by ENTSO-E [134]. This Harmonised Electricity Market Role Model represents agents, their roles, and information flow between them. This role model provides a common definition of roles and domains employed in the electricity market. It enables a common language in the development of information interchange.

Another way to present and describe stakeholders is by using business models. The research by Xia et al. [135] investigates the Swedish mobile phone business ecosystem. The stakeholder listing is represented by the Osterwalder and Pigneur business model canvas. An overview of the agents, their interrelations, and information flows can be illustrated in the business model canvas. The homogenous setup provided by the business model canvas highlights and organizes the identified information. This simplifies the information search. Furthermore, the business model canvas can easily be extended with new stakeholders by following the canvas approach. Both stakeholder listing by diagrams and the business model canvas provide well-organized information about complex systems. The canvas approach makes it possible to include supplementary information about the stakeholders.

5.2. Limitations and Future Work

This paper applies a transparent scoping review methodology through the entire process. To ensure a broad search of the literature, the search strategy includes four online databases, resulting in over

1400 articles. However, the search result may still not identify all relevant articles in the literature despite this paper attempts to be as comprehensive as possible. Ontology is a recently established word in information science [71], therefore, an extension of the literature search including the terms "domain knowledge" and "knowledge representation" may result in additional literature in the field of MAS and ontology.

Furthermore, the fields of organizational theory and business ecosystem are not included in this paper because the literature search only focuses on the energy domain. The energy domain consists of multiple agents and can be considered as an ecosystem in which a community of organisms interacts with each other and the surrounding inorganic environment. This biological definition of an ecosystem is first introduced in [136] and is later adopted in the business domain [137–141]. A business ecosystem is a network of players that are bound together through collective activities to produce an entity that offers value for customers and meet their requirements. MAS in the smart energy domain is similar to this since all types of stakeholders, e.g., electricity traders, building managers, and commercial heat pump providers are connected and interact with each other to offer value for the entire system. Meanwhile, the economic globalization, increasing number of transnational organizations, and rapidly information technology changes increase the complexity of the energy domain, and computational models provide opportunities to understand and respond to these changes [142]. Therefore, a review of the organizational research and business ecosystem in the MAS-orientated energy domain should be considered for future work.

Author Contributions: Conceptualization, M.J.S. and Z.M.; Methodology, Z.M.; writing—Z.M. and M.J.S.; Review and editing, K.C., M.V., Y.D. and B.N.J.; supervision, Z.M., B.N.J. and Y.D.

Funding: This research received no external funding.

Conflicts of Interest: The authors declare no conflict of interest.

Appendix A

Table A1. Selected publication and their focus aspects.

Year	Title	Reference	Focused Aspect		
			Energy Domain	**Ontology**	**MAS Design**
2004	A policy-driven multi-agent system for OGSA-compliant grid control	[124]	Grid control	Policy ontology Application-specific ontology	Agent type Agent logics
2005	Issues in integrating existing multi-agent systems for power engineering applications	[91]	Grid control	Upper ontology Ontology mapping	Inter-MAS communication, Interoperability
2006	Modeling energy and transport infrastructures as a multi-agent system using a generic ontology	[14]	Grid control	Generic and case-specific ontologies	ABM
2007	Multi-agent architecture of energy management system based on IEC 61970 CIM	[13]	Management system	IEC 61970 Standard	Agent structure MAS architecture Multi-agent communication
2009	Multi-agents for energy efficient comfort agents for the energy infrastructure of the built environment: Flexergy	[51]	Buildings/demand side	Ontology for the design process	Agent type
2011	Intelligent multi-agent framework for power system control and protection	[20]	Grid control	Ontology structure	Agent type Agent logics MAS architecture UML diagrams
2011	Multi-agent system for self-optimizing power distribution grids	[15]	Grid control	Domain ontology in the world model	Agent type World model
2013	An architecture for a microgrid-based eco industrial park using a Multi-Agent System	[17]	Microgrid	Ontology in the agent design process	Agent types Agent logics MAS architecture, Negotiation methodology
2013	Demonstration of a multi-agent-based control system for active electric power distribution grids	[23]	Grid control	An ontology with four levels	Agent type MAS architecture
2013	Power transformer condition monitoring and fault diagnosis with multi-agent system based on ontology reasoning	[24]	Grid control	Ontology reasoning	MAS architecture
2013	Upper ontology for multi-agent energy systems' applications	[21]	Power system	Upper ontology and standards	Agent types MAS interoperability
2013	Smart grid - building energy management system: an ontology multi-agent approach to optimize comfort demand and energy supply	[16]	Buildings/demand side	Ontology hierarchical	Agent UML diagrams

Table A1. *Cont.*

Year	Title	Reference	Focused Aspect		
			Energy Domain	Ontology	MAS Design
2014	Energy efficient automation model for office buildings based on ontology, agents and IEC 61499 function blocks	[53]	Buildings/demand side	Translator agent and ontology agent	Agent type Agent logics
2014	Realistic multi-agent simulation of competitive electricity markets	[37]	Electricity market	Upper ontology	Agent types MAS interoperability
2015	Multi-agent simulation of competitive electricity markets: Autonomous systems cooperation for European market modeling	[49]	Electricity market	Upper ontology	MAS interoperability and UML diagrams
2016	Optimal real-time dispatch for integrated energy systems: an ontology-based multi-agent approach	[52]	Grid control	Ontology-based FIPA-ACL	Agent type Communication architecture
2016	Ontology-based demand-side flexibility management in smart grids using a multi-agent system	[50]	Buildings/demand side	Standard of data models in the power system	Gaia methodology
2016	An ontology-driven approach for modeling a multi-agent-based electricity market	[81]	Electricity market	Ontology-Driven Conceptual Modelling	Model-driven development MAS organizational structure
2016	Enabling communications in heterogeneous multi-agent systems: electricity markets ontology	[48]	Electricity market	Electricity Markets OntologyDescription logic	MAS interoperability
2017	A multi-agent-based energy management solution for integrated buildings and microgrid system	[8]	Management systemMicrogrid Buildings/demand side	Ontology for message content	Agent types Agent goals, MAS architecture
2017	EPEX ontology: enhancing agent-based electricity market simulation	[3]	Electricity market	Lower ontology	MAS interoperability
2017	Nord Pool ontology to enhance electricity markets simulation in MASCEM	[47]	Electricity market	Lower ontology	MAS interoperability
2018	Power systems simulation using ontologies to enable the interoperability of multi-agent systems	[38]	Power system	SEAS knowledge model	MAS interoperability
2018	Multi-agent decision support tool to enable interoperability among heterogeneous energy systems	[9]	Power systemMicrogrid	Ontology in Tools Control Center (TOOCC) framework	MAS interoperability

References

1. Koritarov, V.S. Real-world market representation with agents. *IEEE Power Eng. Mag.* **2004**, *2*, 39–46. [CrossRef]
2. Wüstenhagen, R.; Menichetti, E. Strategic choices for renewable energy investment: Conceptual framework and opportunities for further research. *Energy Policy* **2012**, *40*, 1–10. [CrossRef]
3. Santos, G.; Pinto, T.; Praca, I.; Vale, Z. EPEX Ontology: Enhancing Agent-Based Electricity Market Simulation. In Proceedings of the 2017 19th International Conference on Intelligent System Application to Power Systems (ISAP), San Antonio, TX, USA, 17–20 September 2017; pp. 1–6.
4. Lamparter, S.; Ankolekar, A.; Oberle, D.; Studer, R.; Weinhardt, C. Semantic specification and evaluation of bids in web-based markets. *Electron. Commer. Res. Appl.* **2008**, *7*, 313–329. [CrossRef]
5. Yang, S.-Y. Developing an energy-saving and case-based reasoning information agent with Web service and ontology techniques. *Expert Syst. Appl.* **2013**, *40*, 3351–3369. [CrossRef]
6. Zhou, Z.; Chan, W.K.; Chow, J.H. Agent-based simulation of electricity markets: A survey of tools. *Artif. Intell. Rev.* **2007**, *28*, 305–342. [CrossRef]
7. Mays, N.; Pope, C.; Popay, J. Systematically reviewing qualitative and quantitative evidence to inform management and policy-making in the health field. *J. Health Serv. Res. Policy* **2005**, *10*, 6–20. [CrossRef]
8. Anvari-Moghaddam, A.; Rahimi-Kian, A.; Mirian, M.S.; Guerrero, J.M. A multi-agent based energy management solution for integrated buildings and microgrid system. *Appl. Energy* **2017**, *203*, 41–56. (In English) [CrossRef]
9. Teixeira, B.; Pinto, T.; Silva, F.; Santos, G.; Praca, I.; Vale, Z. Multi-Agent Decision Support Tool to Enable Interoperability among Heterogeneous Energy Systems. *Appl. Sci. Basel* **2018**, *8*, 328. [CrossRef]
10. Van der Hoek, W.; Wooldridge, M. Chapter 24 Multi-Agent Systems. In *Foundations of Artificial Intelligence*; van Harmelen, F., Lifschitz, V., Porter, B., Eds.; Elsevier: Amsterdam, The Netherlands, 2008; Volume 3, pp. 887–928.
11. Wooldridge, M.; Jennings, N.R. Intelligent agents: Theory and practice. *Knowl. Eng. Rev.* **1995**, *10*, 115–152. [CrossRef]
12. Russell, S.J.; Norvig, P. *Artificial Intelligence: A Modern Approach*; Pearson Education: London, UK, 2003; p. 1132.
13. Song, W.; Wang, X.; Cao, H.; Pan, G. Multi-Agent Architecture of Energy Management System Based on IEC 61970 CIM. In Proceedings of the 2007 International Power Engineering Conference (IPEC 2007), Singapore, 3–6 Decemeber 2007; pp. 1366–1370.
14. Van Dam, K.H.; Lukszo, Z. Modelling Energy and Transport Infrastructures as a Multi-Agent System Using a Generic Ontology. In Proceedings of the 2006 IEEE International Conference on Systems, Man and Cybernetics, Taipei, Taiwan, 8–11 October 2006; Volume 1, pp. 890–895.
15. Merdan, M.; Lepuschitz, W.; Strasser, T.; Andren, F. Multi-Agent System for Self-Optimizing Power Distribution Grids. In Proceedings of the 5th International Conference on Automation, Robotics and Applications, Wellington, New Zealand, 6–8 December 2011; pp. 312–317.
16. Zeiler, W.; Boxem, G. Smart Grid—Building Energy Management System: An Ontology Multi-agent Approach to Optimize Comfort Demand and Energy Supply. *ASHRAE Trans.* **2013**, *119*, 1–8.
17. Mert, B.; Aradag, U.; Uludag, S.; Unver, H.O. An Architecture for a Microgrid-Based Eco Industrial Park Using a Multi-Agent System. In Proceedings of the 4th International Conference on Power Engineering, Energy and Electrical Drives, Istanbul, Turkey, 13–17 May 2013; pp. 1658–1664.
18. McArthur, S.D.; Davidson, E.M.; Catterson, V.M.; Dimeas, A.L.; Hatziargyriou, N.D.; Ponci, F.; Funabashi, T. Multi-Agent Systems for Power Engineering Applications—Part I: Concepts, Approaches, and Technical Challenges. *IEEE Trans. Power Syst.* **2007**, *22*, 1743–1752. [CrossRef]
19. Jennings, N.R.; Bussmann, S. Agent-based control systems: Why are they suited to engineering complex systems? *IEEE Control. Syst. Mag.* **2003**, *23*, 61–73.
20. Manickam, A.; Kamalasadan, S.; Edwards, D.; Simmons, S. Intelligent Multi-Agent Framework for Power System Control and Protection. In Proceedings of the 2011 IEEE Power and Energy Society General Meeting, San Diego, CA, USA, 24–29 July 2011; pp. 1–8.
21. Santos, G.; Pinto, T.; Vale, Z.; Morais, H.; Praça, I. *Upper Ontology for Multi-Agent Energy Systems' Applications*; Springer International Publishing: Berlin, Germany, 2013; pp. 617–624.

22. Honold, J.; Kandler, C.; Wimmer, P.; Schropp, B.; Reichle, R.; Gröne, M.; Bünemann, M.; Klein, J.; Kufner, M. Distributed integrated energy management systems in residential buildings. *Appl. Therm. Eng.* **2017**, *114*, 1468–1475. [CrossRef]

23. Prostejovsky, A.; Merdan, M.; Schitter, G.; Dimeas, A. Demonstration of a Multi-Agent-Based Control System for Active Electric Power Distribution Grids. In Proceedings of the 2013 IEEE International Workshop on Inteligent Energy Systems (IWIES), Vienna, Austria, 14 November 2013; pp. 76–81.

24. Samirmi, F.D.; Tang, W.; Wu, H. Power Transformer Condition Monitoring and Fault Diagnosis with Multi-Agent System Based on Ontology Reasoning. In Proceedings of the 2013 IEEE PES Asia-Pacific Power and Energy Engineering Conference (APPEEC), Kowloon, China, 8–11 December 2013; pp. 1–6.

25. Chouhan, S.; Wan, H.; Lai, H.J.; Feliachi, A.; Choudhry, M.A. Intelligent Reconfiguration of Smart Distribution Network Using Multi-Agent Technology. In Proceedings of the 2009 IEEE Power & Energy Society General Meeting, Calgary, AB, Canada, 26–30 July 2009; pp. 1–6.

26. Zhang, W.; Xu, Y.; Liu, W.; Zang, C.; Yu, H. Distributed Online Optimal Energy Management for Smart Grids. *IEEE Trans. Ind. Inform.* **2015**, *11*, 717–727. [CrossRef]

27. Pournaras, E.; Vasirani, M.; Kooij, R.E.; Aberer, K. Decentralized Planning of Energy Demand for the Management of Robustness and Discomfort. *IEEE Trans. Ind. Inform.* **2014**, *10*, 2280–2289. [CrossRef]

28. Wang, C. Analysis of Eco-Industrial Park Supporting System on Industrial Symbiosis. In Proceedings of the 2011 IEEE 18th International Conference on Industrial Engineering and Engineering Management, Changchun, China, 3–5 September 2011; Volume 2, pp. 1337–1339.

29. National Institute of Standards and Technology. NIST Framework and Roadmap for Smart Grid Interoperability Standards, Release 3.0, U.S. Department of Commerce 2014. Available online: https://www.nist.gov/publications/nist-framework-and-roadmap-smart-grid-interoperability-standards-release-30 (accessed on 25 July 2019).

30. Morozumi, S. Micro-Grid Demonstration Projects in Japan. In Proceedings of the 2007 Power Conversion Conference—Nagoya, Nagoya, Japan, 2–5 April 2007; pp. 635–642.

31. Lopes, J.A.P.; Moreira, C.L.; Madureira, A.G. Defining control strategies for MicroGrids islanded operation. *IEEE Trans. Power Syst.* **2006**, *21*, 916–924. [CrossRef]

32. Zhao, B.; Zhang, X.; Chen, J. Integrated Microgrid Laboratory System. *IEEE Trans. Power Syst.* **2012**, *27*, 2175–2185. [CrossRef]

33. Dimeas, A.L.; Hatziargyriou, N.D. Operation of a multiagent system for microgrid control. *IEEE Trans. Power Syst.* **2005**, *20*, 1447–1455. [CrossRef]

34. Pipattanasomporn, M.; Rahman, S. Intelligent Distributed Autonomous Power Systems (IDAPS) and Their Impact on Critical Electrical Loads. In Proceedings of the First IEEE International Workshop on Critical Infrastructure Protection (IWCIP'05), Darmstadt, Germany, 3–4 November 2005; p. 9.

35. Khan, A.; Giraud-Audine, C.; Bigot, R.; Abba, G.; Abideen, M.Z. An alternative explanation of forming force reduction for forming process submitted to vibration: Influence of the waveform in the viscoplastic domain. *J. Mater. Process. Technol.* **2016**, *230*, 288–299. [CrossRef]

36. Zhu, Z.; Lambotharan, S.; Chin, W.H.; Fan, Z. A Game Theoretic Optimization Framework for Home Demand Management Incorporating Local Energy Resources. *IEEE Trans. Ind. Inform.* **2015**, *11*, 353–362. [CrossRef]

37. Pinto, T.; Santos, G.; Vale, Z.; Praça, I.; Lopes, F.; Algarvio, H. Realistic Multi-Agent Simulation of Competitive Electricity Markets. In Proceedings of the 25th International Workshop on Database and Expert Systems Applications, Munich, Germany, 1–5 September 2014; pp. 109–113.

38. Santos, G.; Silva, F.; Teixeira, B.; Vale, Z.; Pinto, T. Power Systems Simulation Using Ontologies to Enable the Interoperability of Multi-Agent Systems. In Proceedings of the 2018 Power Systems Computation Conference (PSCC), Dublin, Ireland, 11–15 June 2018; pp. 1–7.

39. Li, H. Development of Open Source Software for Power Market Research: An Illustrative Case Study. *J. En. Mark.* **2009**, *2*, 111.

40. Veselka, T.; Boyd, G.; Conzelmann, G.; Koritarov, V.; Macal, C.; North, M.; Schoepfle, B.; Thimmapuram, P. Simulating the behavior of electricity markets with an agent-based methodology: The Electric Market Complex Adaptive Systems (EMCAS) model. In Proceedings of the USAEE Conference, United States Association for Energy Economic, Houston, TX, USA, 25–27 April 2001.

41. Conzelmann, G.; Boyd, G.; Koritarov, V.; Veselka, T. Multi-Agent Power Market Simulation Using EMCAS. In Proceedings of the IEEE Power Engineering Society General Meeting, San Francisco, CA, USA, 16 June 2005; pp. 2829–2834.

42. Praca, I.; Ramos, C.; Vale, Z.; Cordeiro, M. MASCEM: A multiagent system that simulates competitive electricity markets. *IEEE Intell. Syst.* **2003**, *18*, 54–60. [CrossRef]
43. Praça, I.; Ramos, C.; Vale, Z. Multi-Agent Simulation of Electricity Markets. In Proceedings of the 2nd Workshop Agent-Based Simulation, Boston, MA, USA, July 2000; pp. 89–94.
44. Vale, Z.; Pinto, T.; Praca, I.; Morais, H. MASCEM: Electricity Markets Simulation with Strategic Agents. *IEEE Intell. Syst.* **2011**, *26*, 9–17. [CrossRef]
45. Von Neumann, J.; Morgenstern, O. *Theory of Games and Economic Behavior*; Princeton University Press: Princeton, NJ, USA, 2007.
46. Praça, I.; Ramos, C.; Vale, Z.; Cordeiro, M. Intelligent agents for negotiation and game-based decision support in electricity markets. *Eng. Intell. Syst. Electr. Eng. Commun.* **2005**, *13*, 147.
47. Santos, G.; Pinto, T.; Praça, I.; Vale, Z. Nord Pool Ontology to Enhance Electricity Markets Simulation in MASCEM. In Proceedings of the EPIA Conference on Artificial Intelligence, Porto, Portugal, 5–8 September 2017; Volume 10423, pp. 283–294.
48. Santos, G.; Pinto, T.; Vale, Z.; PraÇA, I.; Morais, H. Enabling Communications in Heterogeneous Multi-Agent Systems: Electricity Markets Ontology. *ADCAIJ Adv. Distrib. Comput. Artif. Intell. J.* **2016**, *5*, 15–42. (In English) [CrossRef]
49. Santos, G.; Pinto, T.; Morais, H.; Sousa, T.M.; Pereira, I.F.; Fernandes, R.; Praça, I.; Vale, Z. Multi-agent simulation of competitive electricity markets: Autonomous systems cooperation for European market modeling. *Energy Convers. Manag.* **2015**, *99*, 387–399. (In English) [CrossRef]
50. Hippolyte, J.L.; Howell, S.; Yuce, B.; Mourshed, M.; Sleiman, H.A.; Vinyals, M.; Vanhée, L. Ontology-Based Demand-Side Flexibility Management in Smart Grids Using a Multi-Agent System. In Proceedings of the 2016 IEEE International Smart Cities Conference (ISC2), Trento, Italy, 12–15 September 2016; pp. 1–7.
51. Zeiler, W.; van Houten, R.; Boxem, G.; van der Velden, J.; Wortel, W.; Haan, J.F.; Noom, P.; Kamphuis, R.; Hommelberg, M.; Broekhuizen, H. Multi-Agents For Energy Efficient Comfort Agents for the Energy Infrastructure of the Built Environment: Flexergy. In Proceedings of the International Conference on Agents and Artificial Intelligence, Icaart 2009, Porto, Portugal, 19–21 January 2009; pp. 579–586.
52. Anvari-Moghaddam, A.; Guerrero, J.M.; Rahimi-Kian, A.; Mirian, M.S. Optimal Real-Time Dispatch for Integrated Energy Systems: An Ontology-Based Multi-Agent Approach. In Proceedings of the IEEE Conferences, Vancouver, BC, Canada, 27–30 June 2016; pp. 1–7.
53. Mousavi, A.; Yang, C.; Pang, C.; Vyatkin, V. Energy Efficient Automation Model for Office Buildings Based on Ontology, Agents and IEC 61499 Function Blocks. In Proceedings of the 2014 IEEE Emerging Technology and Factory Automation (ETFA), Barcelona, Spain, 16–19 September 2014; pp. 1–7.
54. Joyrama, H. A critical evaluation on the factors impacting the adoption of eco-block as a green construction material: From a Mauritian perspective. *J. Build. Eng.* **2019**, *5*, 100789. [CrossRef]
55. Berardi, U. A cross-country comparison of the building energy consumptions and their trends. *Resour. Conserv. Recycl.* **2017**, *123*, 230–241. [CrossRef]
56. Goyal, S.; Wang, W.; Brambley, M.R. Design and implementation of a test bed for building controls. *Build. Serv. Eng. Res.* **2019**. [CrossRef]
57. Rutishauser, U.; Joller, J.; Douglas, R. Control and learning of ambience by an intelligent building. *IEEE Trans. Syst. Man Cybern. Part A Syst.Hum.* **2005**, *35*, 121–132. [CrossRef]
58. Biermann, J.; Corubolo, F.; Eggers, A.; Waddington, S. An Ontology Supporting Planning, Analysis, and Simulation of Evolving Digital Ecosystems. In Proceedings of the 8th International ACM Conference on Management of Digital EcoSystems, Hendaye, France, 2–4 November 2016; pp. 26–33.
59. Georgievski, I.; Degeler, V.; Pagani, G.A.; Nguyen, T.A.; Lazovik, A.; Aiello, M. *Optimizing Offices for the Smart Grid*; Institute for Mathematics and Computer Science University of Groningen: Gorningen, The Netherlands, 2011.
60. Lewis, R. *Modelling Control Systems Using IEC 61499: Applying Function Blocks to Distributed Systems*; The Institution of Engineering and Technology: London, UK, 2001.
61. Lavbič, D.; Vasilecas, O.; Rupnik, R. Ontology-based multi-agent system to support business users and management. *Technol. Econ. Dev. Econ.* **2010**, *16*, 327–347. [CrossRef]
62. Kahn, M.L.; Cicalese, C.D. *The CoABS Grid*; Springer: Berlin, Germany, 2003; pp. 125–134.

63. Steinbrink, C.; Blank-Babazadeh, M.; El-Ama, A.; Holly, S.; Lüers, B.; Nebel-Wenner, M.; Ramírez Acosta, R.P.; Raub, T.; Schwarz, J.S.; Stark, S.; et al. CPES Testing with mosaik: Co-Simulation Planning, Execution and Analysis. *Appl. Sci.* **2019**, *9*, 923. [CrossRef]

64. Hopkinson, K.; Xiaoru, W.; Giovanini, R.; Thorp, J.; Birman, K.; Coury, D. EPOCHS: A platform for agent-based electric power and communication simulation built from commercial off-the-shelf components. *IEEE Trans. Power Syst.* **2006**, *21*, 548–558. [CrossRef]

65. Lin, H.; Veda, S.S.; Shukla, S.S.; Mili, L.; Thorp, J. GECO: Global Event-Driven Co-Simulation Framework for Interconnected Power System and Communication Network. *IEEE Trans. Smart Grid* **2012**, *3*, 1444–1456. [CrossRef]

66. Sun, J.; Tesfatsion, L. An Agent-Based Computational Laboratory for Wholesale Power Market Design. In Proceedings of the 2007 IEEE Power Engineering Society General Meeting, Tampa, FL, USA, 24–28 June 2007; pp. 1–6.

67. Babic, J.; Podobnik, V. *An Analysis of Power Trading Agent Competition*; Springer International Publishing: Cham, Switzerland, 2014; pp. 1–15.

68. Center for Energy Environmental and Economic Systems Analysis. "Electricity Market Complex Adaptive System (EMCAS)", Argonne National Laboratory, Argonne, USA. 2008. Available online: https://ceeesa.es.anl.gov/pubs/61084.pdf (accessed on 20 August 2019).

69. Pinto, T.; Vale, Z.; Praça, I.; Santos, G. *Demonstration of ALBidS: Adaptive Learning Strategic Bidding System*; Springer International Publishing: Cham, Switzerland, 2016; pp. 281–285.

70. Busse, J.; Humm, B.G.; Lübbert, C.; Moelter, F.; Reibold, A.; Rewald, M.; Schlüter, V.; Seiler, B.; Tegtmeier, E.; Zeh, T. Actually, what does "ontology" mean? *J. Comput. Inf. Technol.* **2015**, *23*, 29–41. [CrossRef]

71. Berners-Lee, T.; Hendler, J.; Lassila, O. The Semantic Web. *Sci. Am.* **2001**, *284*, 28–37. [CrossRef]

72. Maedche, A.; Staab, S. Ontology Learning for the Semantic Web. *IEEE Intell Syst.* **2001**, *16*, 72–79.

73. Gruber, T.R. A translation approach to portable ontology specifications. *Knowl. Acquis.* **1993**, *5*, 199–220. [CrossRef]

74. Dillon, T.S.; Chang, E.; Hadzic, M.; Wongthongtham, P. Differentiating Conceptual Modelling from Data Modelling, Knowledge Modelling and Ontology Modelling and a Notation for Ontology Modelling. *APCCM* **2008**, *8*, 7–17.

75. Uschold, M. Knowledge level modelling: Concepts and terminology. *Knowl. Eng. Rev.* **1998**, *13*, 5–29. [CrossRef]

76. Noy, N.F.; McGuinness, D.L. Ontology Development 101: A Guide to Creating Your First Ontology. Stanford University, Stanford. 2001. Available online: https://protege.stanford.edu/publications/ontology_development/ontology101-noy-mcguinness.html (accessed on August 2019).

77. Huang, J.; Dou, D.; He, L.; Hayes, P.; Dang, J. Ontology-Based Knowledge Discovery and Sharing in Bioinformatics and Medical Informatics: A Brief Survey. In Proceedings of the 2010 Seventh International Conference on Fuzzy Systems and Knowledge Discovery, Yantai, China, 10–12 August 2010; Volume 5, pp. 2203–2208.

78. Kim, S. Implementation of Web Ontology for Semantic Web Application. In Proceedings of the Sixth International Conference on Advanced Language Processing and Web Information Technology (ALPIT 2007), Luoyang, Henan, China, 22–24 August 2007; pp. 159–164.

79. Tran, T.; Lewen, H.; Haase, P. On the Role and Application of Ontologies in Information Systems. In Proceedings of the 2007 IEEE International Conference on Research, Innovation and Vision for the Future, Hanoi, Vietnam, 5–9 March 2007; pp. 14–21.

80. Wang, X.H.; Zhang, D.Q.; Gu, T.; Pung, H.K. Ontology Based Context Modeling and Reasoning Using OWL. In Proceedings of the 2004 IEEE Annual Conference on Pervasive Computing and Communications Workshops, Orlando, FL, USA, 14–17 March 2004; pp. 18–22.

81. Poveda, G.; Schumann, R. An Ontology-Driven Approach for Modeling a Multi-Agent-Based Electricity Market. In *German Conference on Multiagent System Technologies*; Springer: Cham, Switzerland, 2016; Volume 9872, pp. 27–40.

82. Tilipakis, N.; Douligeris, C.; Neris, A. Ontology-Based Tools for the Management of Customers Portfolios In a Deregulated Electricity Market Environment. In *Metadata and Semantics*; Springer: Boston, MA, USA, 2009; pp. 269–278.

83. Kofler, M.J.; Reinisch, C.; Kastner, W. A semantic representation of energy-related information in future smart homes. *Energy Build.* **2012**, *47*, 169–179. [CrossRef]

84. Ma, Q.; Yang, Y.; Guo, J. Design of ontology-based decision support software system for grid dispatching. *Autom. Electr. Power Syst.* **2004**, *28*, 57–62.

85. Alexopoulos, P.; Kafentzis, K.; Zoumas, C. ELMO: An Interoperability Ontology for the Electricity Market. In Proceedings of the ICE-B, Milan, Italy, 7–10 July 2009; pp. 15–20.

86. Gruber, T.R. Toward principles for the design of ontologies used for knowledge sharing? *Int. J. Hum. Comput. Stud.* **1995**, *43*, 907–928. [CrossRef]

87. Miller, E. An Introduction to the Resource Description Framework. *Bull. Am. Soc. Inf. Sci. Technol.* **1998**, *25*, 15–19. [CrossRef]

88. Christophides, V. Resource Description Framework (RDF) Schema (RDFS). In *Encyclopedia of Database Systems*; Liu, L., ÖZsu, M.T., Eds.; Springer: Boston, MA, USA, 2009; pp. 2425–2428.

89. Bobillo, F.; Straccia, U. *Uncertainty Reasoning for the Semantic Web III: ISWC International Workshops, URSW 2011–2013, Revised Selected Papers*; Springer International Publishing: Basel, Switzerland, 2014.

90. Baader, F.; Calvanese, D.; McGuinness, D.; Patel-Schneider, P.; Nardi, D. *The Description Logic Handbook: Theory, Implementation, and Applications*; Cambridge University Press: Cambridge, UK, 2003.

91. Catterson, V.M.; Davidson, E.M.; McArthur, S.D.J. Issues in Integrating Existing Multi-Agent Systems for Power Engineering Applications. In Proceedings of the 13th International Conference on, Intelligent Systems Application to Power Systems, Arlington, VA, USA, 6–10 November 2005; p. 6.

92. Hepp, M. Semantic Web and semantic Web services: Father and son or indivisible twins. *IEEE Internet Comput.* **2006**, *10*, 85–88. [CrossRef]

93. Cardoso, J. The Semantic Web Vision: Where Are We? *IEEE Intell. Syst.* **2007**, *22*, 84–88. [CrossRef]

94. Jaeger, M.C.; Rojec-Goldmann, G.; Liebetruth, C.; Mühl, G.; Geihs, K. Ranked Matching for Service Descriptions Using OWL-S. In *Kommunikation in Verteilten Systemen (KiVS) Berlin, Heidelberg 2005*; Springer: Berlin, Germany, 2005; pp. 91–102.

95. De Bruijn, J.; Lausen, H.; Polleres, A.; Fensel, D. The Web Service Modeling Language WSML: An Overview. In *The Semantic Web: Research and Applications, Berlin; 2006*; Springer: Berlin, Germany, 2006; pp. 590–604.

96. Ankolekar, A.; Burstein, M.; Hobbs, J.R.; Lassila, O.; Martin, D.; McDermott, D.; McIlraith, S.A.; Narayanan, S.; Paolucci, M.; Payne, T. DAML-S: Web Service Description for the Semantic Web. In *The Semantic Web—ISWC 2002*; Springer: Berlin, Germany, 2002; pp. 348–363.

97. Kuropka, D.; Tröger, P.; Staab, S.; Weske, M. *Semantic Service Provisioning*; Springer: Berlin, Germany, 2008.

98. Guarino, N. Formal Ontology in Information Systems. In Proceedings of the First International Conference (FOIS'98), Trento, Italy, 6–8 June 1998; IOS Press: Amsterdam, The Netherlands, 1998.

99. Noy, N.F. Semantic integration: A survey of ontology-based approaches. *SIGMOD Rec.* **2004**, *33*, 65–70. [CrossRef]

100. Azevedo, F.; Vale, Z.A.; de Oliveira, P.B. A decision-support system based on particle swarm optimization for multiperiod hedging in electricity markets. *IEEE Trans. Power Syst.* **2007**, *22*, 995–1003. [CrossRef]

101. Pinto, T.; Vale, Z.; Rodrigues, F.; Morais, H.; Praça, I. Bid definition method for electricity markets based on an adaptive multiagent system. In *Advances on Practical Applications of Agents and Multiagent Systems*; Springer: Berlin, Germany, 2011; pp. 309–316.

102. Oliveira, P.; Pinto, T.; Morais, H.; Vale, Z. MASGriP—A Multi-Agent Smart Grid Simulation Platform. In Proceedings of the 2012 IEEE Power and Energy Society General Meeting, San Diego, CA, USA, 22–26 July 2012; pp. 1–8.

103. Van Dam, K.H.; Nikolic, I.; Lukszo, Z. *Agent-Based Modelling of Socio-Technical Systems*; Springer: Dordrecht, The Netherlands, 2012.

104. Kalfoglou, Y.; Schorlemmer, M. Ontology mapping: The state of the art. *Knowl. Eng. Rev.* **2003**, *18*, 1–31. [CrossRef]

105. Zhu, J. Survey on Ontology Mapping. *Phys. Procedia* **2012**, *24*, 1857–1862. [CrossRef]

106. Gomez-Perez, A.; Fernández-López, M.; Corcho, O. *Ontological Engineering: With Examples from the Areas of Knowledge Management, e-Commerce and the Semantic Web. (Advanced Information and Knowledge Processing)*; Springer-Verlag: Berlin, Germany, 2007.

107. Kalyanpur, A.; Parsia, B.; Sirin, E.; Grau, B.C.; Hendler, J. Swoop: A Web Ontology Editing Browser. *J. Web Semant.* **2006**, *4*, 144–153. [CrossRef]

108. Uschold, M.; King, M.; Moralee, S.; Zorgios, Y. The Enterprise Ontology. *Knowl. Eng. Rev.* **1998**, *13*, 31–89. [CrossRef]

109. McArthur, S.D.; Davidson, E.M.; Catterson, V.M.; Dimeas, A.L.; Hatziargyriou, N.D.; Ponci, F.; Funabashi, T. Multi-Agent Systems for Power Engineering Applications—Part II: Technologies, Standards, and Tools for Building Multi-agent Systems. *IEEE Trans. Power Syst.* **2007**, *22*, 1753–1759. [CrossRef]

110. Davidson, E.M.; McArthur, S.D.J.; McDonald, J.R.; Cumming, T.; Watt, I. Applying multi-agent system technology in practice: Automated management and analysis of SCADA and digital fault recorder data. *IEEE Trans. Power Syst.* **2006**, *21*, 559–567. [CrossRef]

111. Wooldridge, M.; Jennings, N.R.; Kinny, D. The Gaia Methodology for Agent-Oriented Analysis and Design. *Auton. Agents Multi Agent Syst.* **2000**, *3*, 285–312. [CrossRef]

112. Hluchý, L.; Laclavik, M.; Balogh, Z.; Babik, M. AgentOWL: Semantic Knowledge Model and Agent Architecture. *Comput. Artif. Intell.* **2006**, *25*, 421–439.

113. Bordini, R.H.; Braubach, L.; Dastani, M.; Seghrouchni, A.E.F.; Gomez-Sanz, J.J.; Leite, J.; O'Hare, G.; Pokahr, A.; Ricci, A. A Survey of Programming Languages and Platforms for Multi-Agent Systems. *Informatica (Slovenia)* **2006**, *30*, 33–44.

114. Herborn, T.; Mondorf, A.; Mougouie, B.; Wimmer, M.A. Semantic Interoperability in the BRITE Project: Ontologies as a Tool for Collaboration, Cooperation and Knowledge Management. In *OTM Confederated International Conferences" On the Move to Meaningful Internet Systems; Monterrey, Mexico, 9–14 November, 2008*; Springer: Berlin, Germany, 2008; pp. 475–483.

115. Sadik, A.R.; Urban, B. An Ontology-Based Approach to Enable Knowledge Representation and Reasoning in Worker–Cobot Agile Manufacturing. *Future Internet* **2017**, *9*, 90. [CrossRef]

116. Poslad, S.; Buckle, P.; Hadingham, R. The FIPA-OS Agent Platform: Open Source for Open Standards. In Proceedings of the 5th International Conference and Exhibition on the Practical Application of Intelligent Agents and Multi-Agents, London, UK, 10–12 April 2000; Volume 355, p. 368.

117. Shostak, I.; Volobuyeva, L.; Danova, M. Ontology Based Approach for Green Software Ecosystem Formalization. In Proceedings of the 2018 IEEE 9th International Conference on Dependable Systems, Services and Technologies (DESSERT), Kiev, Ukraine, 24–27 May 2018; pp. 568–571.

118. Chang, S.K. *Handbook of Software Engineering & Knowledge Engineering*; World Scientific: Singapore, 2001; Volume 2.

119. Vrba, P.; Mařík, V.; Siano, P.; Leitão, P.; Zhabelova, G.; Vyatkin, V.; Strasser, T. A Review of Agent and Service-Oriented Concepts Applied to Intelligent Energy Systems. *IEEE Trans. Ind. Inform.* **2014**, *10*, 1890–1903. (In English) [CrossRef]

120. Chaib-draa, B.; Dignum, F. Trends in agent communication language. *Comput. Intell.* **2002**, *18*, 89–101. [CrossRef]

121. Photovoltaics, D.G.; Storage, E. *IEEE Guide for Smart Grid Interoperability of Energy Technology and Information Technology Operation with the Electric Power System (EPS), End-Use Applications, and Loads, IEEE Std 2030–2011*; IEEE: Piscataway, NJ, USA, 2011; pp. 1–126.

122. Santodomingo, R.; Uslar, M.; Specht, M.; Rohjans, S.; Taylor, G.; Pantea, S.; Bradley, M.; McMorran, A. IEC 61970 for Energy Management System Integration. In *Smart Grid Handbook*; IEEE: Piscataway, NJ, USA, 2016; pp. 1–29.

123. Borgida, A. Description logics in data management. *IEEE Trans. Knowl. Data Eng.* **1995**, *7*, 671–682. [CrossRef]

124. Bei-shui, L.; Ji, G.; Jun, H.; Jiu-jun, C. A Policy-Driven Multi-Agent System for OGSA-Compliant Grid Control. In Proceedings of the 2004 IEEE International Conference on Systems, Man and Cybernetics (IEEE Cat. No.04CH37583), The Hague, The Netherlands, 10–13 October 2004; Volume 6, pp. 5525–5530.

125. FIPA. *FIPA Ontology Service Specification*; FIPA: Geneva, Switzerland, 2000.

126. Uslar, M.; Specht, M.; Rohjans, S.; Trefke, J.; González, J.M. *The Common Information Model. CIM: IEC 61968/61970 and 62325—A Practical Introduction to the CIM*; Springer: Berlin, Germany, 2012.

127. Howell, S.; Rezgui, Y.; Hippolyte, J.-L.; Jayan, B.; Li, H. Towards the next generation of smart grids: Semantic and holonic multi-agent management of distributed energy resources. *Renew. Sustain. Energy Rev.* **2017**, *77*, 193–214. [CrossRef]

128. Guerrero, J.M.; Chandorkar, M.; Lee, T.-L.; Loh, P.C. Advanced control architectures for intelligent microgrids—Part I: Decentralized and hierarchical control. *IEEE Trans. Ind. Electron.* **2013**, *60*, 1254–1262. [CrossRef]

129. Santodomingo, R.; Rohjans, S.; Uslar, M.; Rodriguez-Mondejar, J.; Sanz-Bobi, M.A. Ontology matching system for future energy smart grids. *Eng. Appl. Artif. Intell.* **2014**, *32*, 242–257. [CrossRef]

130. Papasratorn, B.; Charoenkitkarn, N.; Vanijja, V.; Chongsuphajaisiddhi, V. *Proceedings of the Advances in Information Technology: 6th International Conference IAIT 2013, 12–13 December*; Springer International Publishing: Bangkok, Thailand, 2013.

131. Luncean, L.; Becheru, A. Communication and Interaction in a Multi-Agent System Devised for Transport Brokering. In Proceedings of the 2015 Balkan Conference on Informatics: Advances in ICT, Craiova, Romania, 2–4 September 2015; Volume 1427, pp. 51–58.

132. Man, D. Ontologies In Computer Science. *Didact. Math.* **2013**, *31*, 43–46.

133. Dong, H.; Hussain, F.K.; Chang, E. Ontology-Based Digital Ecosystem Conceptual Representation. In Proceedings of theThird International Conference on Autonomic and Autonomous Systems (ICAS'07), Athens, Greece, 19–25 June 2007; p. 42.

134. ENTSO-E. *The Harmonised Electricity Role Model*; ENTSO: Brussels, Belgium, 2018.

135. Xia, R.; Rost, M.; Holmquist, L.E. Business Models in the Mobile Ecosystem. In Proceedings of the 2010 Ninth International Conference on Mobile Business and 2010 Ninth Global Mobility Roundtable (ICMB-GMR), Athens, Greece, 13–15 June 2010; pp. 1–8.

136. Willis, A.J. The ecosystem: An evolving concept viewed historically. *JSTOR* **1997**, *11*, 268–271.

137. Moore, J.F. Predators and prey: A New Ecology of Competition. In *Creating Value in THE Network Economy*; Harvard Business School Press: Brighton, MA, USA, 1999; pp. 121–141.

138. Dong, H.; Hussain, F.K. Digital Ecosystem Ontology. In Proceedings of the 2007 IEEE Conference on Emerging Technologies and Factory Automation (EFTA 2007), Vigo, Spain, 4–7 June 2007; pp. 814–817.

139. Burlacu, G.; Cojocaru, L.E.; Danila, C.; Popescu, D.; Stanescu, A.M. A Digital Business Ecosystem Integrated Approach for Farm Management Information System. In Proceedings of the 2nd International Conference on Systems and Computer Science, Villeneuve d'Ascq, France, 26–27 August 2013; pp. 80–85.

140. Pappas, N.; Kazasis, F.G.; Anestis, G.; Gioldasis, N.; Christodoulakis, S. A Knowledge Management Platform for Supporting Digital Business Ecosystems based on P2P and SOA technologies. In Proceedings of the 2007 Inaugural IEEE-IES Digital EcoSystems and Technologies Conference, Cairns, Australia, 21–23 February 2007; pp. 196–202.

141. Peltoniemi, M.; Vuori, E. Business Ecosystem as the New Approach to Complex Adaptive Business Environments. In *Proceedings of eBusiness Research Forum*; Citeseer: Princeton, NJ, USA, 2008.

142. Prietula, M.; Carley, K.; Gasser, L. *Simulating Organizations: Computational Models of Institutions and Groups*; The MIT Press: Cambridge, MA, USA, 1998.

Article

Variable Renewable Energy and Market Design: New Products and a Real-World Study

Hugo Algarvio [1,2], Fernando Lopes [1,*], António Couto [1], Ana Estanqueiro [1] and João Santana [2]

[1] LNEG–National Laboratory of Energy and Geology, Est. Paço Lumiar 22, 1649-038 Lisbon, Portugal; hugo.algarvio@tecnico.ulisboa.pt (H.A.); antonio.couto@lneg.pt (A.C.); ana.estanqueiro@lneg.pt (A.E.)
[2] Instituto Superior Técnico, 1049-001 Lisbon, Portugal; jsantana@ist.utl.pt
* Correspondence: fernando.lopes@lneg.pt; Tel.: +351-210-924-600

Received: 14 October 2019; Accepted: 27 November 2019; Published: 30 November 2019

Abstract: Most existing energy markets (EMs) were not designed to take into account an active participation of variable renewable energy (VRE). This situation results typically in imbalances and substantial costs in balancing markets. Such costs are reflected both in the energy and the VRE parts of the consumer tariffs. Both appropriate market products and new elements of market design may largely facilitate the large-scale integration of VRE in EMs. Accordingly, this article presents a new bilateral energy contract and introduces two new marketplaces that can contribute to reduce the imbalances resulting from VRE producers. It also presents a study conducted with the help of an agent-based tool, called MATREM. The results indicate a significant decrease in the imbalances and the associated costs.

Keywords: day-ahead market; balancing market; bilateral trading; market design; variable renewable energy; agent-based simulation; MATREM system

1. Introduction

Energy markets (EMs) are a complex and continuously evolving reality, meaning that new players are emerging—chief among these are the producers of variable renewable energy (VRE)—and new challenges need to be managed—such as the ones associated with the participation of VRE producers in competitive markets [1,2]. Indeed, recent years have witnessed a substantial increase of non-controllable or variable renewable energy, notably wind power and solar photovoltaic. VRE has several unique characteristics compared to those of conventional generation, including significant fixed capital costs but near-zero or zero production costs. VRE is also normally the marginal resource, since it is operated at maximum capacity (taking into account the weather conditions). These characteristics have a strong influence on the outcomes of EMs, reducing market-clearing prices [3]. Accordingly, the research community has paid attention to the effectiveness of current market designs to determine if they are still efficient to deal with the increasing levels of VRE (see, e.g., [4,5]).

VRE typically involves significant forecast errors, which may result in large imbalances. The day-ahead market (DAM) closes normally at 12:00 p.m. (CET), and thus the bids of wind power producers need to be calculated by taking into account power forecasts computed 12 to 36 h ahead. As a result, an adjustment of the gate closure to a time closer to real-time operation seems to be important to enable a fair participation of VRE producers. The differences between the quantities of energy produced and the commitments resulting from the DAM need to be balanced in the intra-day market and/or the balancing market. At present, the participation of VRE producers in balancing markets (BMs) is still very limited, despite the technical feasibility and the (potential) motivation to operationalize such participation.

To address the issues associated with the participation of VRE producers in markets, adaptations to the current market structure as well as new elements of market design have been proposed by theorists and practitioners working on the area of competitive energy markets. For instance, the International Energy Agency points out that the physical transactions of electrical energy in power systems with high shares of VRE need to be made by considering auctions and centralized pools, and should not take into account feed-in-tariffs or other supporting schemes. The process of trading energy also needs to be improved by defining the terms of the transactions up to 30 min before real-time operation with an interval up to 10 min [6]. This near real-time negotiation is also supported by the Clean Energy Package (Article 7), published by the European Commission [7]. In this package, a new proposal for regulating the Internal Market for Electricity is presented, with the main goals of stimulating the global leadership of Europe in renewables, harmonizing markets rules, supporting the integration of VRE, and increasing the general welfare of consumers (see [8] for a complete overview). Article 6 of the new proposal indicates that market operators should develop new products to accommodate the increasing levels of VRE and support demand-response programs.

Now, generally speaking, European markets typically allow bidding up to 5 to 30 min before real-time operation, contributing to reduce the imbalances resulting from VRE producers. The markets of North-America and Australia present some additional flexibility by including 5-min real-time (sub-)markets. Despite this, most real-world markets operate by considering power (MW) and not energy (MWh), thus allowing to some extent substantial deviations of VRE producers.

Against this background, this article presents a new bilateral energy contract, called short-term energy (STE) contract, and introduces two new marketplaces that may allow to reduce the imbalances resulting from VRE producers (we note that throughout the article the terms "new marketplace" and "new market product" will be used interchangeably). The main aim is to enable an active and competitive participation of VRE producers in energy markets, decreasing imbalances and the associated costs, and to some extent avoiding the waste of energy. The new contract and the design of the new marketplaces take into account the following aspects: legal basis, market time unit, minimum bid quantity, transaction time horizon, type of market participant, and the role of participants in the process of trading energy. The authors are aware of no similar market products in place in the real-world.

Furthermore, the article presents a simulation-based study to analyze the behavior (and test) the new contract and the design of the new market places in a real-world setting. The study involves the participation of both wind power producers (WPPs) and retailers in markets, who prepare bids according to different strategies. The simulations are performed with the help of the agent-based tool called MATREM (see [9,10]). Six key performance indicators (KPIs) are considered, namely the value of wind energy to the market, the global imbalances of the system, the imbalances and costs of WPPs and retailers, and the total cost of the system.

The work presented here builds on our previous work in the areas of trading wind power in markets [11,12] and portfolio optimization of retailers [13]. Specifically, in [11], we investigated the benefits of the participation of WPPs in BMs at both economic and technical levels. In [12], we analyzed the impact of the wind power forecast uncertainty and the change of the day-ahead market gate closure on market outcomes. In [13], we introduced a model for optimizing the portfolios of retailers using the Markowitz theory. In this paper, as noted, we present and test a contract and two marketplaces related to the participation of VRE producers in energy markets.

The remainder of the paper is structured as follows. Section 2 presents the main features of existing energy markets. Section 3 discusses the participation of VRE producers in balancing markets. Section 4 presents the new energy contract and Section 5 the new marketplaces. Section 6 summarizes the features of the MATREM system. Section 7 illustrates the trading behavior of WPPs by taking into account the new contract and marketplaces. Section 8 presents the simulation-based study and discusses the experimental results. Finally, Section 9 presents some concluding remarks.

2. Energy Markets and VRE Producers

Day-ahead markets close typically at 12:00 p.m., 12 to 36 h before physical delivery. Market participants trade energy on exchanges or pools using programs based on the system marginal pricing theory. Prices and quantities are calculated in a specific day D for every hour of day $D + 1$. Intra-day markets are essentially markets involving scheduling and pricing procedures a few hours ahead to facilitate balancing in advance of real-time. Such markets may involve various sessions based on auctions or may operate continuously (see, e.g., [14]). Most American markets also include a short term market, generally referred to a real-time market, to set prices and schedules for 5-min periods (but see [3]).

Derivatives are financial instruments that include forwards, futures, options and swaps [15]. These instruments are essentially contracts to buy or sell a specific amount of electricity at a certain future time for a specific price. They may span from days to several years and allow market participants to hedge against the financial risk inherent to day-ahead and intra-day prices [16]. Also, they may be financial (involving a purely financial settlement) or physical (involving a financial settlement and the physical delivery of energy), and are typically traded in derivatives exchanges. In short, market participants submit orders to sell or buy electricity in an electronic trading platform. Orders include the quantity and the price as well as several other parameters that are deemed appropriate. The trading platform automatically and continuously matches the orders that are likely to interfere with each other (typically, for a particular type of contract and a specific energy price). Also, apart from derivatives exchanges, bilateral contracts—such as forwards and swaps—may be negotiated privately between two parties. The terms of such contracts are very flexible and can be defined to meet the objectives and needs of both parties (but see [17] for a more in-depth discussion).

Balancing markets are imposed by the European Network of transmission system operators and allow to compensate the deviations from the schedules defined in day-ahead and intra-day markets, as well as in bilateral contracts. The players that deviate typically need to pay penalties. The system operators have access to reserve capacity for the provision of system services, namely primary reserve (or frequency control reserve), secondary reserve (or fast active disturbance reserve), and tertiary reserve (or slow active disturbance reserve). Primary reserve is the first to be activated, after grid disturbances or imbalances between production and consumption. It must be activated up to 15 s and the disturbances need to be controlled in 30 s. Secondary reserve should be fully activated in 30 s and can continue active for a maximum of 15 min. Tertiary reserve is activated manually, up to 15 min, and can continue active for hours (see, e.g., [8]).

Secondary and tertiary reserve are traded by system operators in day-ahead tenders. In short, these agents define the needs of the power system for up and down-regulation, receive the proposals of the authorized participants, and determine schedules and prices by using an algorithm based on the system marginal pricing theory. Typically, different simulations are performed for computing the price for up and down-regulation. Now, apart from bilateral contracts and derivatives exchanges, balancing markets are most important for the work described here, and the next section is devoted to the participation of VRE producers in such markets.

3. Participation of VRE Producers in Balancing Markets

3.1. Status of some European Countries and Product Analysis

Considering the technical feasibility of the participation of wind power producers in balancing markets, several authors acknowledge this possibility, in case the current market rules and product specifications are adapted (see, e.g., [18–20]). In this way, and although with some restrictions, notably the fact that WPPs need to prepare bids aggregated with conventional generation, Spain [18], Germany [19] and Denmark [20] have already allowed the participation of WPPs in BMs (although for downward regulation only).

Great Britain allowed WPPs to participate in two curtailment products, namely "manage constraint" and "rebalance system", receiving 40% more money to curtail energy than to produce it, which is often not considered an efficient way to use VRE (but see [21]). In Belgium, some researchers studied the participation of WPPs in BMs, considering the downward automatic-activated frequency restoration reserve (aFRR), obtaining a reliability higher than 90% (see [22]). And for the case of EU-28, a study considering the participation of WPPs, solar producers and other renewable energy producers in BMs, indicated a reduction of 6% in the costs associated with such markets [23].

Now, considering existing and emerging market products associated with VRE producers and BMs, the provision of reactive power may be considered an important product [24]. Also, primary reserve is a potential product for VRE producers, contributing to compensate the disadvantage of a reduced inertia in power systems with high levels of VRE [25]. Photovoltaic systems do not have mass inertia and can adjust their output within milliseconds. Wind turbines can deliver primary reserve faster than is currently required. Also, wind turbines can deliver synthetic inertia, which can solve the problem of the reduced inertia of the grid due to high shares of VRE [26]. In this way, the well-known ramping products [27] and the P2X solutions [28] are important aspects to explore. We note, however, that fast ramping schedules in frequency control ancillary services (FCAS) can avoid the curtailment of VRE, although some limitations prevent an adequate participation of VRE producers (e.g., fixed sloping schedules). Also, the prices associated with power to X solutions (P2X), such as power-to-hydrogen, are currently very low.

All of the mentioned "products" share a common shortcoming: they contribute to an increase of the waste of energy or curtailment of VRE, by making VRE producers participating in non-optimal schedules. Furthermore, all of them consider the technical capabilities of VRE, instead of the optimal use of VRE without curtailments, wasting energy and allowing a large (short-term) interaction between market participants, notably VRE producers, retailers and transmission system operators (TSOs).

3.2. Bid Preparation and Energy Deviations

The participation of VRE producers in balancing markets involves the preparation and submission of bids, typically for a period of one hour, creating large difficulties for producers to keep a stable and efficient operation without curtailments or deviations. Specifically, WPPs need to bid their expected active power, $P_{bid}(T)$, in day-ahead and intra-day markets, based on forecasts that use time horizons between 18 h and 42 h (for DAMs), and between 2 h and 7 h (for intra-day markets), ahead of real-time operation. At real-time, considering the data obtained from wind parks, there is the possibility to compute the average production, $P_{avg}(T)$, for a particular period of time T.

From the point of view of the grid, WPPs produce instantaneous power, $p(t)$, based on the wind speed, the characteristics and power of wind turbines and the configuration of wind parks. The energy, $E(T)$, produced during a period of time T, is computed as follows:

$$E(T) = \int_0^T p(t)dt = T \times P_{avg}.$$ (1)

The deviation, $E_{dev}(T)$, in period T is given by the following formula:

$$E_{dev}(T) = T \times P_{avg} - T \times P_{bid} = T(P_{avg} - P_{bid}).$$ (2)

Figure 1 illustrates the relation between these variables for a particular hour of operation ($T = 1$). It shows the traded/scheduled power, $P_{bid}(T)$, of a particular wind power producer in the day-ahead market (red curve). The corresponding energy, $E_{bid}(T)$, is represented by the orange area. The real production of the wind power producer is given by $p(t)$ (solid blue curve). The difference between the instantaneous power and the scheduled power gives the energy deviation, $E_{dev}(T)$, which is represented by the light blue area.

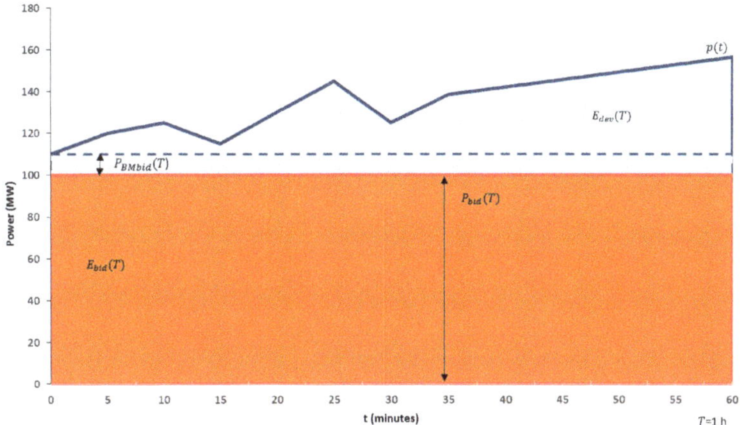

Figure 1. Bids of wind power producers and the associated deviations.

Considering now that wind power producers are allowed to make only single bids in the balancing market (i.e., bids without considering aggregation of WPPs), the maximum expected power that they can submit, $P_{BMbid}(T)$, is defined by the difference between the minimum instantaneous deviation and the traded power:

$$P_{BMbid}(T) = (-1)^c \times (min\,(|p(t)|) - P_{bid}(T))$$

(3)

where $t = 1, \ldots, 60$, and the bid is of the type upward regulation, i.e., $c = 0$ (for a downward regulation bid, consider the maximum instead of the minimum, and $c = 1$).

Figure 1 shows that the submission of $P_{BMbid}(T)$ to the balancing market, and its subsequent acceptance by the system operator, results in a waste of energy (the difference between the dashed and solid blue curves). For the particular case of the secondary reserve market, the waste of energy may be even larger than that. Accordingly, existing products associated with the balancing market may not be considered adequate to deal with the variability and uncertainty of VRE. A possible solution to overcome the problem is to consider aggregated bids involving VRE producers and conventional generation. However, not all power systems allow aggregated bids, specially for the case of BMs.

4. The Short-Term Energy Contract

The short-term energy (STE) contract is a new type of bilateral contract—and to some extent a new market product—that presents some similarities with the aforementioned bilateral contracts, although there are obvious differences. The specifications of the contract are shown in Table 1. It has the goal of allowing agents to reduce/avoid imbalances and consequently the potential payment of penalties—that is, it is not a profit-seeking product. Accordingly, the energy price is pre-defined as the market-clearing price (DAM price) for the period under consideration. The minimal energy quantity is 0.1 MWh. Agents submit to a trading platform bids involving specific energy quantities for periods of 15 min (and not one hour). Bids may be new or associated with energy deviations and should be submitted up to one hour prior to real-time operation. Buy and sell bids likely to interfere with each other generate transactions and new contracts. To this end, the trading platform takes into account energy quantities only (quantities may be either fully or partially matched). Bids associated with deviations have priority over new bids and are matched according to a principle of equity. Physical delivery is done in strict accordance with transmission system operators, who are informed about the terms and conditions of new contracts (energy price, energy quantity, etc.).

Table 1. Main specifications of the short-term energy contract.

Characteristic	Details
Trading procedure/period	Auction/year-round
Time unit	15 min (of each day)
Description	Electronic trading managed by exchanges; agents offer their energy deviations in exchanges, informing TSOs of deals
Match type	Full or partial
Key feature	Different contracts for positive and negative energy
Bid submission	Till 1 h prior to the balancing period
Key bid elements	Energy quantity, match type, etc.
Energy quantity	Minimum of 0.1 MWh; traded quantity based on an equity principle
Type of bid	Bid associated with an energy deviation or a new bid; deviations have priority over new bids
Payment scheme	Day-ahead price
Participants	VRE producers, conventional generation producers, retailers

Now, an important feature—and to the best of our knowledge—a novel feature of the STE contract is the inherent aspect of considering energy and not power. Figure 2 illustrates this aspect by depicting bids involving either power (green line) or energy (blue line). Bids are assumed to be simple and consist of quantities and other parameters that are deemed appropriate (e.g., match type). The settlement period has the duration of 15 min. Existing bilateral contracts consider typically a quantity based on power, meaning that power plants should follow a constant production schedule during the settlement period (green line of Figure 2, corresponding to a quantity of 50 MW). This may not be adequate for wind power producers and other VRE producers due to the uncertainty and variability of renewable generation. Accordingly, the SET contract considers a quantity based on energy—that is, power plants do not necessarily need to follow a constant production schedule during the settlement period (orange line of Figure 2, corresponding to an average quantity of 12.5 MWh). This typically leads to a decrease of the imbalances and the associated costs (but see the real-world study presented in Section 8). Overall, despite the existence of a number of contracts traded in energy markets worldwide, such as daily future contracts or even 15-min base and peak contracts (see, e.g., [29,30]), it is especially noteworthy that we are aware of no contracts similar to the short-term energy contract. At this stage, we note that an appealing alternative to the short-term energy contract involves the submission of both energy price and energy quantity (instead of energy quantity only). However, energy would be traded at different prices and, to some extent, some transactions would not be considered due to the mismatch of price.

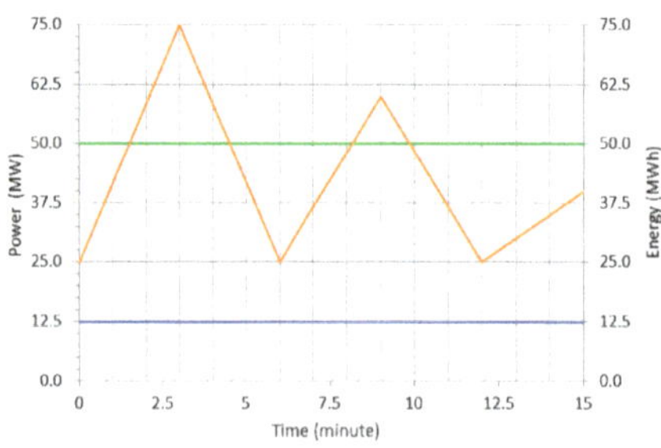

Figure 2. Real production of a hypothetical variable renewable energy (VRE) producer (orange line) and bids of power (green line) and energy (blue line) for a settlement period of 15 min.

Table 2. Important features of the renewable power band marketplace.

Characteristic	Details
Trading procedure/period	Auction/year-round
Market time unit	15 min (of each day)
Description	VRE producers use their frequency control capacity at balancing markets by offering a power band that can help system operators solving demand-supply imbalances
Key feature	Power interval
Bid submission	Till 15 min prior to the balancing period
Key bid elements	Positive and negative power
Bid quantity	Minimum of 0.1 MW; Power interval
Payment scheme	Selected producers receive the secondary reserve price
Participants	VRE producers

5. Trading Reserve Capacity

5.1. Renewable Power Band Marketplace

As noted earlier, the work published in [11] analysed the benefits of the participation of wind power producers in balancing markets at both economic and technical levels. We found that a reduction of the market time unit from 1 h to 15 min is beneficial to WPPs. Accordingly, this section considers a market time unit of 15 min—that is, WPPs submit bids for periods of 15 min, and not for periods of 1 h. In other words, capacity reserve is traded for 15-min periods. To this end, we also consider that VRE producers participate in this marketplace by using the frequency control capacity, offering a power band for each 15-min period, and thus participating with scarce real-time instantaneous power. The upper limit should be lower than the expected optimal power, and the lower limit should be higher than the technical capacity of VRE producers to reduce generation from an optimal (or intermediary) level to a lower one. Also, bids may involve either positive or negative power intervals and not necessarily intervals ranging from negative to positive values, as typically happens in existing secondary reserve markets. Market participants are allowed to submit bids till 15 min prior to real-time operation. They are remunerated by the secondary reserve price. Table 2 presents some important features of this marketplace. Figure 3 illustrates the trading behavior of VRE producers, by showing the different bids and the associated deviations, for a period of one hour.

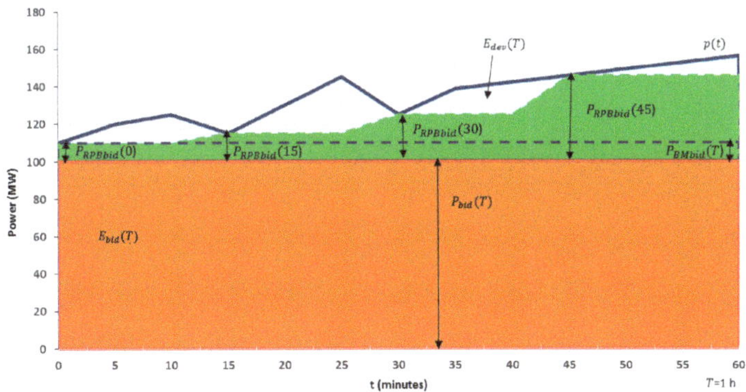

Figure 3. Bids associated with an hypothetical VRE producer for a period of one hour (four periods of 15 min).

Figure 3 is, to some extent, similar to Figure 1, which shows the traded/scheduled power, $P_{bid}(T)$, the corresponding energy, $E_{bid}(T)$, the real production of a wind power producer, $p(t)$, and the energy deviation, $E_{dev}(T)$. In addition, Figure 3 also shows the lack of deviations resulting from the participation in this marketplace—that is, the deviations that are avoided by WPPs—represented by the green area (see also Section 8).

VRE producers should guarantee that they are able to comply with their bids (in order to avoid the payment of penalties). Accordingly, for each 15-min period, the maximum power that they can bid, P_{RPBbid}, is given by the difference between the minimum deviation and the traded bid (see Equation (3)). The wasted energy associated with the accepted bids is equal to the area between the solid blue curve and the dashed green curve (see Figure 3).

5.2. Energy Reserve Marketplace

Again, as indicated in the previous subsection, the market time unit is assumed to be 15 min—that is, VRE producers submit bids for periods of 15 min. Specifically, WPPs offer their energy (MWh) at a near-zero price—that is, bids involve energy, instead of power—and the other market participants typically submit bids at higher prices. All bids are ranked with increasing price (merit-order). The transmission system operator, a non-commercial organization, independent of commercial players, constitutes the demand side. In case this agent is procuring up regulation, the up-regulation bids with lowest prices are activated until the procured quantity is reached. The price of the last up-regulated quantity sets the up-regulation price. The bids with prices below the up-regulation price have a profit, equal to the difference between the final regulation price and the offered price. A similar procedure is used to find the down-regulation price. Players are allowed to submit bids till 15 min prior to real-time operation. They are remunerated by the market-clearing price (tertiary reserve price). Table 3 presents some important features of this marketplace.

Table 3. Important features of the energy reserve marketplace.

Characteristic	Details
Trading procedure/period	Auction/year-round
Market time unit	15 min (of each day)
Description	VRE producers submit their deviations to balancing markets and inform the system operators of the magnitude of errors
Key feature	Different products for positive and negative reserve
Bid submission	Till 15 min prior to the balancing period
Key bid element	Energy
Energy quantity	Minimum of 0.1 MWh
Energy price	Near-zero price
Payment scheme	Selected producers receive the tertiary reserve price
Participants	VRE producers

Overall, VRE producers submit their deviations (MWh) to this marketplace and inform TSOs of the magnitude of the errors, thus avoiding to some extent the payment of large penalties (but see Section 8). It is worth noting that to increase competition in the market VRE producers may submit bids including a higher price (i.e., a price higher than a near-zero price). Specifically, for up-regulation, the price may be equal to the price of the first ranked offer (minimal price). For down-regulation, the price could be defined in a similar way (i.e., the price of the first ranked offer, or maximal price).

6. Main Features of the MATREM System

MATREM (for Multi-Agent TRading in Electricity Markets) is an agent-based tool for simulating the behavior of competitive energy markets. In [9], we present a detailed description of the system, and in [10] we classify the system according to a number of dimensions associated with both electricity markets and intelligent agents. The remainder of this section gives an overview of MATREM.

The system supports a day-ahead market (DAM), an intra-day market (IDM), a futures market, and a balancing market (BM). The DAM is a central market where generation and demand are traded on an hourly basis [31]. The IDM is a short-term market that involves several auction sessions. Both markets operate according to the marginal pricing theory and are controlled by a market operator agent. Two pricing mechanisms are supported: system marginal pricing (SMP) and locational marginal pricing (LMP). The futures market is a market to hedge against the financial risk (i.e., the price volatility) associated with the DAM and the IDM. It is an organized market for both financial and physical products, which may span from days to years. The balancing market is a market for the provision of system services. MATREM considers three types of reserve, namely primary reserve, secondary reserve and tertiary reserve. The stability of the power system is a task associated with a system operator agent, who is responsible for the operation of this market.

The system also supports a marketplace for negotiating tailored (or customized) bilateral contracts, notably contracts defined to cover the delivery of large amounts of electrical energy over long periods of time. Two types of contracts are considered: forward contracts and contracts for difference [32]. The negotiating parties are equipped with a model that handles two-party and multi-issue negotiation. The negotiation process is an iterative process involving an exchange of offers and counter-offers [33–35].

Market participants are modeled as software agents and include generating companies, retailers, aggregators, traditional consumers, coalitions of consumers, market operators and system operators. VRE aggregators allow the participation of WPPs and other VRE producers in the aforementioned markets. Coalitions of consumers are essentially alliances of end-use customers with the main goal of increasing their bargaining power. The system platform is a 32/64-bit computer running the JAVA programming language and the JADE platform [36].

7. Trading Behavior of Wind Power Producers

As noted earlier, most existing energy markets, including the Iberian electricity market [37], were not designed to deal with large levels of non-dispatchable generation. MIBEL includes a day-ahead market and an intra-day market, managed by the Spanish electricity market operator (OMIE [38]), as well as a derivatives market, managed by the Portuguese electricity market operator (OMIP [39]). Market participants submit hourly bids to trade energy in the day-ahead market. The IDM involves six auction sessions, with gate-closures between 1 and 5 h ahead of real-time operation. The derivatives market allows private parties to trade standardized bilateral contracts.

Figure 4 shows the typical trading behavior of WPPs in the DAM and the IDM, which takes into consideration the most reliable meteorological data, in order to minimize deviations. WPPs submit wind power forecasts in the day-ahead market during day $D-1$, and make commitments to produce specific quantities of energy for day D. The bids to submit to the intra-day market are essentially deviations—that is, they are computed by taking into consideration the actual commitments and the current updates of the wind power forecasts.

In this work, the forecasts are obtained by considering the Neural Network Toolbox and the artificial neuronal networks (ANN) approach [40]. ANN is typically organized in layers, namely the input, hidden, and output layers. Each layer covers one or more nodes that represent the basic unit of information process (also known as the neuron). The main advantage of the ANN is the capability to deal with nonlinear relationships by learning the association between the predictors and predictions.

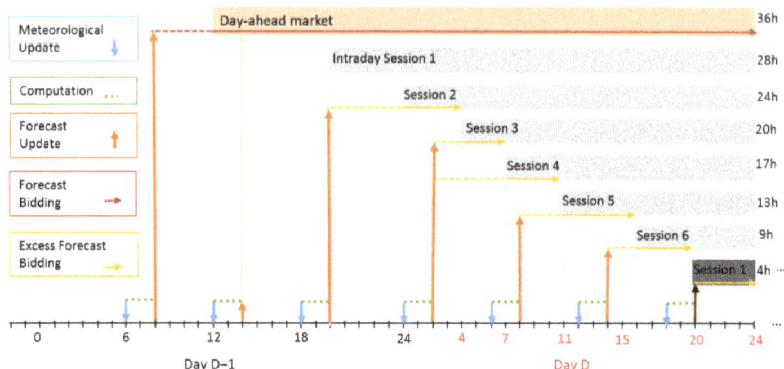

Figure 4. Trading behavior of wind power producers in the day-ahead and intra-day markets.

Numerical weather prediction (NWP) models present systematic errors, which are mainly associated with the following: (i) initial and boundary meteorological conditions (ICs), (ii) a poor representation of the physics of the models, and (iii) a failure to solve sub-grid scale phenomena, such as the sea-land interaction. The data to feed the NWP model to enable wind power producers to participate in the day-ahead market are based on ICs from 06:00 UTC, representing a time horizon for the meteorological forecast ranging between 18 to 42 h ahead. For the case of the intra-day market, the time horizon for the forecast ranges from 2 to 7 h (depending on the auction session).

The meteorological data (ICs) are updated every six hours, namely at 6:00, 12:00, 18:00 and 24:00 (see the small light-blue arrows of Figure 4). This data needs to be processed to get the wind power forecasts, involving a computation time of nearly 2 h (see the dotted green lines of Figure 4). The updated forecasts are represented by the vertical orange arrows, meaning that the bids of WPPs to intra-day sessions 1, 3, 5 and 6 are based on updated data, and the bids to the other sessions take into account the available data only. Also, the bids of WPPs for a particular intra-day session involve the first hours of that session only—that is, the hours until the beginning of the next session (see the horizontal dark yellow arrows of Figure 4). Such bids include a quantity and a price. The quantity refers to the excess of energy. The price is 0 €/MWh.

Overall, WPPs submit bids to the day-ahead market, make commitments to produce energy during day *D*, and submit their deviations to the various sessions of the intra-day market. After that, in case WPPs still expect deviations, they can make use of the new bilateral energy contract to eventually reduce the imbalances. Following this, WPPs can participate in the aforementioned marketplaces. At the end, in case WPPs still expect deviations, they need to assume their balance responsibility and (eventually) pay penalties for their imbalances.

8. Simulation-Based Study

This section presents a simulation-based study carried out by using the MATREM system. It involves the simulation of the day-ahead and intra-day markets, as well as the use of the new contract and the simulation of the new marketplaces.

8.1. Data, Agents and Scenarios

The following sources of data are considered: (i) hourly prices and quantities of the day-ahead and intra-day Iberian markets (data published by the Spanish electricity market operator, OMIE [38]), (ii) hourly prices and quantities submitted to the Portuguese balancing market (data reported by the Portuguese system operator, REN [41]), and (iii) hourly deviations and prices of the imbalances for producers and retailers (data reported by REN [41]). Also, the study makes use of real wind power data from a set of wind parks located in the central region of Portugal, which is available from 1 January 2009 to 31 December 2010.

The (software) agents are 12 producers (with several production units), representing the supply-side of Portugal, five retailers, representing the demand-side of Portugal and Spain, and one aggregated wind power producer, with 249 MW of installed capacity (representing 10% of the Portuguese installed capacity in 2010). To get expressive results, the data was upscaled to 2490 MW of installed capacity, by multiplying all values by a constant factor.

Table 4 presents several key features of the producer agents. As noted earlier, the forecasts for the day-ahead and intra-day markets (time horizon ranging from 18 h to 42 h, in case of the day-ahead market, and from 2 to 7 h, for the intra-day market), were obtained by considering a numerical weather prediction model coupled with an artificial neuronal networks approach. Both the numerical prediction model and the artificial neuronal approach were calibrated for the region under consideration. The normalized root mean square error of the day-ahead forecast is around 13.5%. For the new products, the wind power forecasts were obtained using the historical time series and the ANN approach.

Table 4. Key features of the producer agents.

Agent Identifier	Iberian Country	Generation Technology	Maximum Capacity (MW)	Marginal Cost (€/MWh)
P1	Portugal	Wind	2500	0
P2	Portugal	Renewable mix	2000	0
P3	Portugal	Hydro	4500	$[30, 60]$
P4	Portugal	Coal	1800	≈ 30
P5	Portugal	Gas Combined Cycle	3000	≈ 55
P6	Portugal	Fuel oil	2000	≈ 70
P7	Spain	Renewable mix	30,000	0
P8	Spain	Hydro	16,500	$[30, 60]$
P9	Spain	Coal	10,000	≈ 30
P10	Spain	Nuclear	7500	≈ 30
P11	Spain	Gas Combined Cycle	22,000	≈ 55
P12	Spain	Fuel oil	4000	≈ 70

The study involves the following seven scenarios:

- S1: Wind power producers participate in the day-ahead market only (baseline scenario);
- S2: WPPs participate in both the day-ahead and the intra-day markets;
- S3: WPPs participate in the DAM and the IDM, and make use of the STE contract; also, retailers and conventional generation participate in the DAM and the IDM, and make use of the STE contract;
- S4: WPPs participate in the day-ahead and intra-day markets, as well in the energy reserve marketplace;
- S5: WPPs participate in the DAM and the IDM, as well in the renewable power band marketplace;
- S6: WPPs participate in the DAM, the IDM, and the renewable power band and energy reserve marketplaces;
- S7: WPPs participate in the DAM, the IDM, use the STE contract, and participate in the renewable power band and energy reserve marketplaces; also, retailers and conventional generation participate in the DAM and the IDM, and use the STE contract.

For all scenarios, we consider that the agents assume their balance responsibility, including the wind power producers, meaning that all agents pay penalties for their imbalances. To quantify the relevance of both the new contract and the new marketplaces, several important parameters are simulated, notably:

- Deviations of wind power producers and retailers;
- Deviation costs of wind power producers and retailers;
- Wind energy value to the market.

8.2. Results and Discussion

Tables 5 and 6, and Figures 5 and 6 summarize the results obtained by simulating the behavior of the day-ahead and intra-day markets with the help of the MATREM system, as well as considering the new contract and the new marketplaces. In particular, Table 5 shows the main results for the wind power producers, namely the energy value, the deviations and the remunerations. The table indicates that the STE contract (scenario S3) substantially reduced the deviations of the wind power producers, namely 56% in relation to the base scenario (see also Figure 5). In this way, the use of this contract decreased the cost of the deviations of WPPs by nearly 59% (see also Figure 6). However, WPPs were remunerated by the day-ahead clearing price, which was normally less attractive than the price of the tertiary reserve market (mFRR market). The participation of WPPs in the energy reserve marketplace (scenario S4) also increased the wind energy value (16% in relation to the base scenario S1). This result is explained by the high remuneration received by WPPs in the mFRR market, for their excess of energy (according to the actual operation of the Iberian market). In relation to the participation of WPPs in the renewable power band marketplace (scenario S5), they obtained a high levelized remuneration from energy. However, in comparison with the other scenarios, WPPs trade a lower quantity of energy, which results in a lower remuneration, and a higher cost with deviations.

Table 5. Simulation results for the wind power producers (energy value, deviations and remunerations).

Scenario	Energy Value (€/MWh)	Deviation (MW/h)	Deviation Cost (k€/h)	Remuneration (Energy) (€/MWh)	Remuneration (Power) (€/MW)	Liquid Profit (€/MWh)
S1	28.05	281.53	4.54	—	—	—
S2	29.51	210.05	3.13	—	—	—
S3	31.29	123.10	1.83	−15.00	—	14.99
S4	32.58	148.61	2.00	−7.14	—	36.48
S5	31.57	195.83	2.89	64.26	29.06	25.07
S6	34.31	142.00	1.88	−2.50	29.06	34.61
S7	33.25	98.66	1.40	−11.52	28.83	18.74

Table 6. Simulation results for the retailer agents.

Scenario	Deviation (MW/h)	Deviation Cost (k€/h)	Remuneration (Energy) (€/MWh)	Liquid Profit (€/MWh)
S1	—	—	—	—
S2	431.91	6.07	—	—
S3	262.67	3.75	18.14	5.24
S4	—	—	—	—
S5	—	—	—	—
S6	—	—	—	—
S7	262.67	3.75	18.14	5.24

Table 5 indicates that the use of the STE contract and the participation of WPPs in the energy reserve marketplace (scenarios S3 and S4) are particularly important, compensating their real-time lack of energy (i.e., wind power forecast underestimations). An analysis of the fifth column of the table shows that WPPs need to buy energy to fulfill their commitments with the DAM and the IDM. For the case of scenario S5, involving the participation of WPPs in the renewable power band marketplace, the benefits are not substantial for situations involving wind power forecast underestimations. In such situations, WPPs receive money for their excess of energy, instead of paying for their lack of energy.

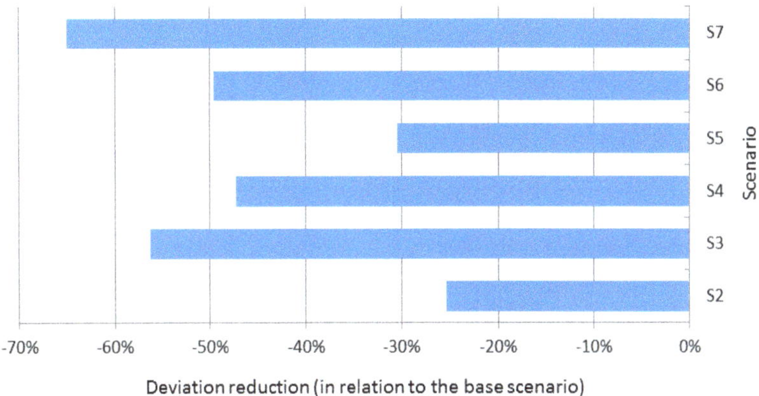

Figure 5. Simulation results (reduction of deviations of the wind power producers).

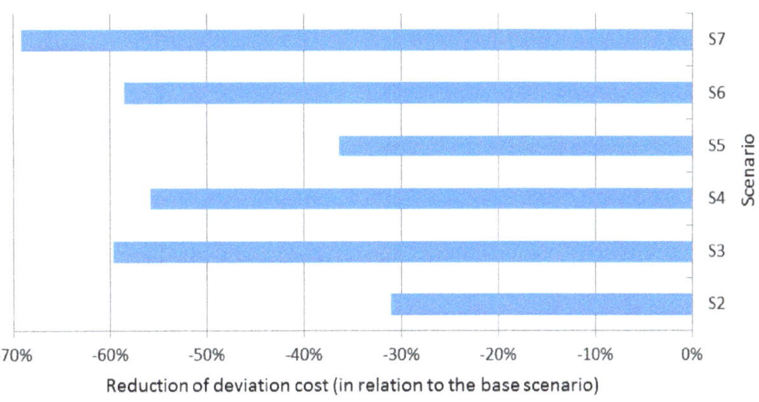

Figure 6. Simulation results (reduction of deviation cost of wind power producers).

The sixth column of the table reflects to some extent the price of the secondary reserve market (aFRR market). WPPs receive a higher remuneration for power (scenario S5, corresponding to 29.06 €/MW), in comparison with the remuneration resulting from scenario S7 (28.83 €/MW), involving the new contract and the participation in the new marketplaces. This result is associated with a reduction in deviations that is not attractive, leading to a reduced power band (and bid).

Overall, the results of Table 5 show that the use of the new market products are beneficial to wind power producers. Accordingly, both market operators and independent system operators should take them into consideration, possibly with particular adaptations and/or extensions, in order to allow an effective participation of VRE producers in liberalized markets—that is, to provide the flexibility needed to integrate the increasing levels of renewable generation, maintaining the security of the power systems.

Table 6 shows the simulation results for the particular case of retailers. These agents use the STE contract only. The results indicate that the retailer agents reduce the deviations in 39% and the costs with deviations in 38%. Also, the results for scenario S7 show that the use of the STE contract and the participation in the new marketplaces by WPPs, as well as the use of the new contract by retailers, result in a significant decrease in the global imbalances (-49%) and the associated costs (-51%).

9. Conclusions

This article presented a new bilateral energy contract (STE contract) and new marketplaces designed to allow wind power producers to (eventually) reduce their deviations. The contract and the marketplaces were to some extent tested in a real-world study conducted with the help of the multi-agent system MATREM.

The software agents were 12 producers, representing the supply-side of Portugal, five retailers, representing the demand-side of Portugal and Spain, and one aggregated wind power producer. The data involved hourly prices and quantities of the day-ahead and intra-day Iberian markets, hourly prices and quantities submitted to the Portuguese balancing market, and hourly deviations and prices of the imbalances for producers and retailers. Seven scenarios were considered, involving the participation of wind power producers in the new marketplaces, and also using the new contract.

The simulation results indicated that the new contract and the new marketplaces are beneficial to WPPs at both technical and economical levels, reducing the imbalances and increasing the wind energy value to the market. Also, the STE contract led to a reduction in the imbalances of retailers, resulting in a reduction of their costs with imbalances. Accordingly, market operators and system operators should take into consideration the new products presented here, possibly with particular adaptations and/or extensions, to allow an effective participation of VRE producers in liberalized markets, thus contributing to a new paradigm involving no feed-in tariffs nor other supporting schemes—that is, the paradigm expected in the near future, involving near 100% renewables.

Author Contributions: H.A. and A.C. defined the new contract and the new marketplaces, performed the simulation-based study, and wrote a preliminary version of the article. In particular, A.C. developed the forecast models and computed the wind power forecasts. H.A. designed and conducted the simulations with the help of the MATREM system. F.L. performed a deep revision of the initial version of the article, regarding both language and scientific content. F.L. also performed a deep revision of the subsequent versions of the article, taking into account the comments of the reviewers. In addition, F.L. developed the MATREM system (thus enabling to perform the simulation-based study). A.C. and A.E. contributed with expertise in wind power. J.S. and F.L. contributed with expertise in energy markets.

Funding: This work was supported by FCT (Fundação para a Ciência e Tecnologia) under grant agreement PD/BD/105863/2014 (H. Algarvio).

Abbreviations

aFRR	Automatic-activated frequency restoration reserve
ANN	Artificial neural networks
BM	Balancing market
CG	Conventional generation
CET	Central European time
DAM	Day-ahead market
EM	Electricity market
EU-28	The European Union before Brexit
GFS	Global Forecast System
IC	Initial and boundary conditions
IDM	Intra-day market
KPI	Key performance indicator
LMP	Locational marginal pricing
mFRR	Manually-activated frequency restoration reserve
MATREM	Multi-Agent TRading in Electricity Markets
MIBEL	Iberian electricity market
NWP	Numerical weather prediction
OMIE	Spanish electricity market operator
OMIP	Portuguese electricity market operator

REN	Portuguese TSO
STE	Short-term energy contract
SMP	System marginal pricing
UTC	Coordinated universal time
VRE	Variable renewable energy
TSO	Transmission system operator
WPP	Wind power producer

Indices

t	Time (minutes)
T	Time period (hour)

Parameters

p^{nom}	Nominal capacity

Variables

c	Parameter of Equation (3)
D	Day
$E(T)$	Energy
$E_{bid}(T)$	Energy bid
$E_{dev}(T)$	Deviation
$p(t)$	Instantaneous power
$P_{avg}(T)$	Average power
$P_{bid}(T)$	Power bid
$P_{BMbid}(T)$	Balancing market bid

References

1. Jones, E.L. *Renewable Energy Integration: Practical Management of Variability, Uncertainty, and Flexibility in Power Grids*; Academic Press: London, UK, 2014.
2. Lopes, F.; Coelho, H. *Electricity Markets with Increasing Levels of Renewable Generation: Structure, Operation, Agent-Based Simulation and Emerging Designs*; Springer International Publishing: Cham, Switzerland, 2018. [CrossRef]
3. Ela, E.; Milligan, M.; Bloom, A.; Cochran, J.; Botterud, A.; Townsend, A.; Levin, T. Overview of Wholesale Electricity Markets. In *Electricity Markets with Increasing Levels of Renewable Generation: Structure, Operation, Agent-Based Simulation and Emerging Designs*; Springer International Publishing: Cham, Switzerland, 2018; pp. 3–21. [CrossRef]
4. Moiseeva, E.; Wogrin, S.; Hesamzadeh, M.R. Generation flexibility in ramp rates: Strategic behavior and lessons for electricity market design. *Eur. J. Oper. Res.* **2017**, *261*, 755–771. [CrossRef]
5. Ela, E.; Milligan, M.; Bloom, A.; Botterud, A.; Townsend, A.; Levin, T. Incentivizing Flexibility in System Operations. In *Electricity Markets with Increasing Levels of Renewable Generation: Structure, Operation, Agent-Based Simulation, and Emerging Designs*; Springer International Publishing: Cham, Switzerland, 2018; pp. 95–127. [CrossRef]
6. International Energy Agency. *The Power of Transformation: Wind, Sun and the Economics of Flexible Power Systems*; International Energy Agency: Paris, France, 2014.
7. European Commission. Proposal for a Regulation of the European Parliament and of the Council on the Internal Market for Electricity. February 2017. Available online: http://ec.europa.eu/energy/sites/ener/files/documents/1_en_act_part1_v9.pdf (accessed on 4 October 2019).
8. Algarvio, H.; Lopes, F.; Couto, A.; Santana, J.; Estanqueiro, A. Effects of Regulating the European Internal Market on the integration of Variable Renewable Energy. *WIREs Energy Environ.* **2019**, e346. [CrossRef]
9. Lopes, F. MATREM: An Agent-based Simulation Tool for Electricity Markets. In *Electricity Markets with Increasing Levels of Renewable Generation: Structure, Operation, Agent-Based Simulation and Emerging Designs*; Springer: Cham, Switzerland, 2018; pp. 189–225. [CrossRef]
10. Lopes, F.; Coelho, H. Electricity Markets and Intelligent Agents. Part II: Agent Architectures and Capabilities. In *Electricity Markets with Increasing Levels of Renewable Generation: Structure, Operation, Agent-Based Simulation and Emerging Designs*; Springer: Cham, Switzerland, 2018; pp. 49–77. [CrossRef]

11. Algarvio, H.; Lopes, F.; Couto, A.; Estanqueiro, A. Participation of Wind Power Producers in Day-ahead and Balancing Markets: An Overview and a Simulation-based Study. *WIREs Energy Environ.* **2019**, e343. [CrossRef]

12. Algarvio, H.; Couto, A.; Lopes, F.; Estanqueiro, A. Changing the Day-Ahead Gate Closure to Wind Power Integration: A Simulation-Based Study. *Energies* **2019**, *12*, 2765. [CrossRef]

13. Algarvio, H.; Lopes, F.; Sousa, J.; Lagarto, J. Multi-agent electricity markets Retailer portfolio optimization using Markowitz theory. *Electr. Power Syst. Res.* **2017**, *148C*, 282–294. [CrossRef]

14. Kirschen, D.; Strbac, G. *Fundamentals of Power System Economics*; Wiley: Chichester, UK, 2004.

15. Hull, J. *Options, Futures, and Other Derivatives*; Pearson Education: New York, NY, USA, 2015.

16. Clewlow, L.; Strickland, C. *Energy Derivatives: Pricing and Risk Management*; Lacima Publications: London, UK, 1998.

17. Lopes, F. Electricity Markets and Intelligent Agents. Part I: Market Architecture and Structure. In *Electricity Markets with Increasing Levels of Renewable Generation: Structure, Operation, Agent-Based Simulation and Emerging Designs*; Springer: Cham, Switzerland, 2018; pp. 23–48. [CrossRef]

18. Fernandes, C.; Frías, P.; Reneses, J. Participation of intermittent renewable generators in balancing mechanisms: A closer look into the Spanish market design. *Renew. Energy* **2016**, *89*, 305–316. [CrossRef]

19. Ocker, F.; Ehrhart, K. The "German Paradox" in the balancing power markets. *Renew. Sustain. Energy Rev.* **2017**, *67*, 892–898. [CrossRef]

20. Sorknaes, P.; Andersen, A.; Tang, J.; Strøm, S. Market integration of wind power in electricity system balancing. *Energy Strategy Rev.* **2013**, *1*, 174–180. [CrossRef]

21. National Grid. Monthly Balancing Services Summary 2017/18, Electricity System Operator (ESO) in Great Britain; Technical Report; 2 February 2018. Available online: https://www.nationalgrideso.com/document/109401/download (accessed on 4 October 2019).

22. WindVision; Elia; Eneco; ENERCON. Delivery of Downward aFRR by Wind Farms; Technical Report; Belgium 2015. Available online: https://www.elia.be/ (accessed on 10 November 2019).

23. Fournié, L.; Andrey, C.; Hentschel, J.; Wilkinson, G. *Integration of Electricity Balancing Markets and Regional Procurement of Balancing Reserves*; Final Report; European Commission: Brussels, Belgium, 2016.

24. European Commission. Regulation (EU) 2016/631 establishing a network code on requirements for grid connection of generators. *Off. J. Eur. Union* **2016**. Available online: https://eur-lex.europa.eu/legal-content/EN/TXT/?uri=CELEX%3A32016R0631 (accessed on 1 October 2019)

25. Ren, H.; Zhang, X.; Kang, S.; Liang, S. Actuator Disc Approach of Wind Turbine Wake Simulation Considering Balance of Turbulence Kinetic Energy. *Energies* **2019**, *12*, 16. [CrossRef]

26. Altin, M.; Kuhlmann, J.; Das, K.; Hansen, A. Optimization of Synthetic Inertial Response from Wind Power Plants. *Energies* **2018**, *11*, 1051. [CrossRef]

27. Hu, J.; Wen, F.; Wang, K.; Huang, Y.; Salam, M. Simultaneous provision of flexible ramping product and demand relief by interruptible loads considering economic incentives. *Energies* **2018**, *11*, 46. [CrossRef]

28. Wang, F.C.; Hsiao, Y.S.; Yang, Y.Z. The Optimization of Hybrid Power Systems with Renewable Energy and Hydrogen Generation. *Energies* **2018**, *11*, 1948. [CrossRef]

29. OMIP. *Trading RuleBook*; MIBEL Derivatives Market: Lisbon, Portugal, 2016.

30. EPEX Spot. *Trading on EPEX SPOT*; EPEX SPOT SE: Paris, France, 2018.

31. Algarvio, H.; Couto, A.; Lopes, F.; Estanqueiro, A.; Santana, J. Multi-Agent Energy Markets with High Levels of Renewable Generation: A Case-Study on Forecast Uncertainty and Market Closing Time. In Proceedings of the 13th International Conference Distributed Computing and Artificial Intelligence, Sevilla, Spain, 1–3 June 2016; pp. 339–347.

32. Sousa, F.; Lopes, F.; Santana, J. Contracts for Difference and Risk Management in Multi-agent Energy Markets. In *Advances in Practical Applications of Agents, Multi-Agent Systems, and Sustainability: The PAAMS Collection (PAAMS 2015)*; Springer International Publishing: New York, NY, USA, 2015; pp. 339–347. [CrossRef]

33. Lopes, F.; Mamede, N.; Novais, A.Q.; Coelho, H. Negotiation Tactics for Autonomous Agents, In Proceedings of the 12th International Workshop on Database and Expert Systems Applications (DEXA), Munich, Germany, 3–7 September 2001; pp. 1–5.

34. Lopes, F.; Mamede, N.; Novais, A.Q.; Coelho, H. Negotiation in a multi-agent supply chain system. In *Third Int. Workshop of the IFIP WG 5.7 Special Interest Group on "Advanced Techniques in Production Planning & Control"*; Firenze University Press: Firenze, Italy, 2002; pp. 153–168.

35. Lopes, F; Coelho, H. Concession Behaviour in Automated Negotiation. In *E-Commerce and Web Technologies*; Springer: Berlin/Heidelberg, Germany, 2010; pp. 184–194.

36. Bellifemine, F.; Caire, G.; Greenwood, D. *Developing Multi-agent Systems with JADE*; John Wiley & Sons: Chichester, UK, 2007.

37. MIBEL. Iberian Electricity Market. Available online: https://www.mibel.com/en/home_en/ (accessed on 10 November 2019).

38. OMIE. *Day-Ahead and Intraday Electricity Market Operating Rules*; Spanish Electricity Market Operator: Spain, 2018; pp. 1–186.

39. OMIP. *Trading RuleBook, MIBEL Derivatives Market*; Portuguese Electricity Market Operator: Lisbon, Portugal, 2016; pp. 1–47.

40. MathWorks. *MATLAB and Neural Network Toolbox Release*; The MathWorks, Inc.: Natick, MA, USA, 2017.

41. REN. Redes Elétricas Nacionais, Daily Summary. Available online: http//www.mercado.ren.pt/EN/Electr /MarketInfo/MarketResults/Pages/default.aspx (accessed on 10 November 2019)

Article

Dynamic Tariff for Day-Ahead Congestion Management in Agent-Based LV Distribution Networks

Niyam Haque [1,*], Anuradha Tomar [1], Phuong Nguyen [1,2] and Guus Pemen [1]

[1] Department of Electrical Engineering, Eindhoven University of Technology, 5600 MB Eindhoven, The Netherlands; a.tomar@tue.nl (A.T.); p.nguyen.hong@tue.nl (P.N.); a.j.m.pemen@tue.nl (G.P.)

[2] Sustainable Energy Systems Group, SUSTAIN—ERIN, Luxembourg Institute of Science and Technology, L-4422 Belvaux, Luxembourg

[*] Correspondence: a.n.m.m.haque@tue.nl; Tel.: +31-40-247-8515

Received: 1 November 2019; Accepted: 6 January 2020; Published: 9 January 2020

Abstract: Capacity challenges are becoming more frequent phenomena in residential distribution networks with new forms of loads, distributed renewable energy resources (RES) and price-responsive applications. Different types of demand response programs have been introduced to tackle these challenges through iterative changes in price and/or contractual participations based on incentives. In this research, a dynamic network tariff-based demand response program is proposed to address congestion problems in low-voltage (LV) networks. The formulation takes advantage of the scalable architecture of the agent-based systems that allows local decision making with limited communication. Energy consumption schedules are developed on a day-ahead basis depending on the expected cost of overloading for a number of probable scenarios. The performance of the proposed approach has been tested through simulations in the unbalanced IEEE European LV test feeder. Simulation results reveal up to 82% reduction in congestion on a monthly basis, while maintaining the quality of supply in the network.

Keywords: congestion management; dynamic tariff; agent-based distribution networks; demand response

---

1. Introduction

Fundamental changes are expected in the power systems in terms of increasing shares of renewable energy resources (RES)-based local distributed generation (DG) units and price-responsive residential flexible loads. Due to the intermittent nature of the DG units and reduced diversity of the loads, more and more operational challenges are appearing in residential LV networks in terms of voltage variations, thermal overloading and power quality issues [1]. Congestion or thermal overloading result from power flows through a network asset (lines, cables, transformers) exceeding its transfer capability. Even though the network assets are usually designed to withstand power flows beyond a defined margin, continuous overloading leads to ageing and subsequent failure of the distribution cables and transformer windings [2,3]. The conventional approach of reinforcing the network assets to tackle such issues necessitates a huge investment, and it is also deemed redundant, as the frequency of such issues is unpredictable [4]. In order to avoid the huge costs, different types of demand response (DR) mechanisms have been studied and developed worldwide for the procurement of flexibility from end-users in order to solve these operational challenges. These mechanisms can be roughly subdivided into direct and indirect load control methods. In direct methods, end-users are compensated through an appropriate contract, which enables automatic or manual shift in consumption of electricity based on the requirements of the network operator. Indirect methods, on the other hand, invoke flexibility through a market-based coordination and/or dynamic price signals [5,6]. An overview of such mechanisms has been presented in [7–12] for various types of smart grid domains such as, for instance, microgrids, load balancing services, smart electric vehicle (EV) charging and residential energy management.

A centralized coordinated home energy management system, which implements daily power based network tariffs to mitigate network congestion due to the high demands of heat pumps (HPs) and EVs, is described in [13]. In contrast to conventional centralized methods, a number of decentralized approaches have been proposed in order to adequately address the comfort of the end-users for scheduling the energy consumption of dynamic inputs [14–17]. Such decentralized approaches aim to enable the end-users to take autonomous actions with no or limited communication with other entities. Contrary to the centralized approaches, they offer a more robust and reliable operation due to the lower vulnerability for communication failure.

A local market based framework has been presented in [18] that supports the network operators and prosumers to fully utilize the available flexibility and, thus, enhance the operational efficiency of the network. Dynamic tariff schemes that consider the high level penetration of EV and HP have been discussed in [19–21]. In order to tackle congestion in distribution networks, a dynamic subsidy (DS) method is presented in [22] to motivate aggregators to adjust the load profiles and to take network constraints into consideration. DS is inherently an incentive or reward that the distribution system operator (DSO) is willing to pay to the aggregators who would reschedule their energy profiles. Sensitivity analyses and linear approximation methods have been applied in [23] to develop a pricing model, considering EVs as a flexible load and accounting for three-phase and single-phase voltage variation, network losses and overloading of network components. For the identification of barriers, potential solutions and opportunities, a generic method is applied in the Netherlands [24]. The applied method focuses on market integration aspects and tries to access the flexibility procurement mechanism through the aggregators. As an outcome of the study, an analysis on relevant issues for market integration with reference to the Dutch system and European Grid codes is presented and a lack of proper communication system between users is identified. Authors of [25] have performed an analysis to understand the network pricing methodology most suitable for prosumers in the future, and they concluded that contracts would be economically efficient when retailers have access to the full range of devices, which helps them to convert wholesale electricity pricing in retail contracts, and second, the wholesale price associated with customers must represent the short term marginal cost of electricity production.

However, for a market-based DR mechanism, a sound methodology for congestion management is important to reflect the realized ageing of the network assets to a representative financial loss. An integrated congestion management mechanism has been proposed in [2,16] for the residential networks that involve the dynamic loading model of a distribution transformer. In reality, procurement of flexibility can be a more complex problem involving multiple involved actors in the same network. This work extends the market-based control proposed in [16] for day-ahead scheduling of the residential appliances that considers the dynamic network tariff to avoid thermal overloading of the MV/LV transformer. The tool developed in [16] has been used for estimating the overloading cost for different scenarios of imminent day-ahead loading, and a more realistic approach has been adopted of changing the network tariff dynamically for various scenarios of loading. Thus, the new approach becomes robust for uncertainties associated with day-ahead scheduling. The method will take advantages of Multi-Agent Systems (MAS) for a scalable architecture and distributed intelligence in a multi-actor setting. In principle, the method allows for local decision making and dynamic scheduling of the appliances with limited communication and a smaller number of iterations among the actors. Key contributions of this article can thus be listed as follows:

- Day-ahead market-based congestion management through agent-based scheduling of the residential appliances;

- The estimated incurred cost of overloading is determined through the dynamic loading model of the transformer and used as the trigger for the procurement of flexibility;
- The dynamic network tariff has been calculated through a robust formulation considering uncertainties in loading and correlation among loading in different time steps.

The rest of this paper is organized as follows: Section 2 highlights the overall approach and system architecture; Section 3 presents the problem formulation along with the description of the proposed methodology; Section 4 provides the description of the simulation test case and the assumptions adopted. Finally, simulation results are discussed in Section 5, before summarizing and concluding with Section 6.

2. Approach

2.1. Overview

With the increasing adoption of flexible domestic appliances and small-scale generation technologies like rooftop solar PV, market-based control of the LV network has been drawing extensive research attention in recent years [26–28]. This has principally paved the way for new market actors (e.g., aggregators, energy service companies), novel business models, innovative control algorithms and in general, a more decentralized operation of the future power system [4].

Transactive energy frameworks have been developed to incorporate the flexibility of residential appliances with a range of scopes and aims [29,30]. In this context, market-based mechanisms with local flexibility services have attracted notable interest for ancillary services and network congestion management [6]. However, in a market-coordinated network with a considerable penetration of price-responsive appliances, the overall peak loads tend to be significantly higher due to the higher coincidence of loads in low-price hours. These peak loads may lead to more frequent occurrences of network congestions in the LV networks.

In this work, we aim to incorporate a variable network tariff with the day-ahead dynamic price, as discussed in [6]. The market-based control is coordinated by a commercial market actor such as an aggregator. Aggregators are usually responsible for managing the flexibility in their cluster of end-users and providing the DSO with a schedule of the loading for the following day. Case congestion is expected, and the dynamic price levels are adjusted by varying the network tariff to shift the flexible demand in order to relieve the congestion. Conventional market-based mechanisms try to adjust the day-ahead price based on the network or market issues. These processes need to be carried out by market actors, like aggregators or energy suppliers, who need to be compensated for the provided services and the adjustment in the portfolio. On the contrary, the proposed approach focus on the network tariff part of the price and does not have to depend on the market parties.

The application of such an approach relies on robust and reliable communication among different entities in the whole value chain. An MAS-based system architecture is adopted in this work that provides for an efficient platform for smooth coordination among the involved actors. An overview of the distributed system architecture and related interactions are presented in the following subsection.

2.2. MAS-Based System Architecture

A hierarchical MAS-based architecture is adopted in this work (as shown in Figure 1), consisting of device agents, household agents, an aggregator agent and a network agent. The connected loads and DG units are represented by device agents (DAs). The DAs are responsible for the energy optimization of respective appliances based on the given price signals and associated constraints. DAs of each house are coordinated by a house agent (HA) that interfaces with the external market and network agents through an aggregator. The aggregator coordinates the end-users through dynamic prices and supports the DSO with the required flexibility during network issues.

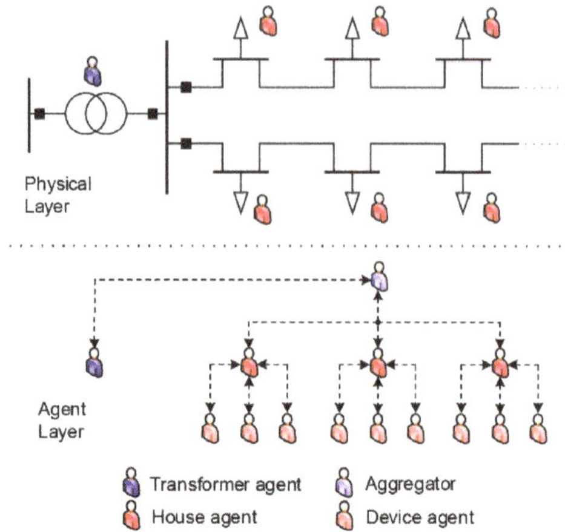

Figure 1. Multi-Agent Systems (MAS)-based system architecture.

The transformer agent (TA) represents a network agent that monitors and controls the MV/LV transformer. The agent-based architecture enables the system to be sufficiently scalable, since more end-users and their appliances can be easily included in the system in terms of new agents. At the same time, the platform can be expanded to wider network areas by implementing more transformers in terms of additional TAs. A more detailed description of the system architecture can be found in [16].

The interactions among the agents in the MAS environment are depicted in Figure 2. In normal operations, the HA adjusts overall energy consumption by optimizing the predicted consumption of the DAs. The final day-ahead profile is sent to the TA to check network security constraints. In case the TA detects an imminent overloading of the transformer, the tariff-based DR phase is triggered (as depicted in Figure 2b). At this phase, the TA calculates the resulting ageing and thermal overloading cost. Based on the cost, it determines required changes in the network tariff at different time instants of the day while keeping the daily average constant. The updated network tariff is sent to the HAs as a flexibility request through the aggregator for an iterative solution of the problem.

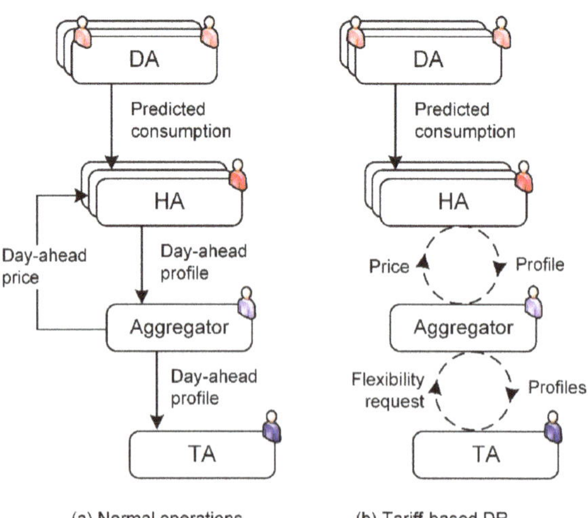

Figure 2. Interaction among the involved agents.

3. Modeling

In this section, the modeling of the proposed mechanism is discussed in detail. At any given time, the total available flexibility depends upon the type of loads in the network. Therefore, in order to procure the required flexibility, the behavior of the loads for different price levels needs to be taken into consideration. The Markov Chain Monte-Carlo method is applied in this regard to perform the modeling of household loads [31].

The day-ahead scheduling of the appliances has two aspects. First, the response of the devices needs to be modeled for dynamic prices, and then their behavior needs to be adjusted for changes in the price if flexibility is requested by the TA.

3.1. Response to Dynamic Pricing

The household appliances can be broadly categorized into four groups, i.e., non-controllable, buffer appliances, time shifting and curtailable appliances [32]. The non-controllable appliances are mostly commonly known as the base load and include the devices that do not offer controllability for different levels of price. Buffer appliances represent devices with integrated storage facilities in terms of thermal (e.g., freezers, heat pump) or electro-chemical energy (e.g., batteries, EV). Time-shifting appliances can shift their consumption in time based on price and include the washing machine, dishwasher, etc. Finally, the curtailable appliances refer to the devices that can be interrupted based on predefined contracts for different purposes (e.g., solar PV system for voltage violation or network congestion). Each type of these appliances would react differently to the incoming price levels. The HA tries to optimize the energy cost of the house by utilizing the available flexible loads, and the associated optimization problem could be expressed as:

$$\min \sum_{a=1}^{|A|} \sum_{t=1}^{N_T} p_t \times P_{t,a} \times \Delta t \tag{1}$$

where A is the set of all appliances within a single household, N_T is the number of time steps in a day, p_t is the price at time t, $P_{t,a}$ is the power of appliance a at time t and Δt is the duration of each time step.

The buffer and time shifting appliances need to optimize their profiles based on the dynamic price. Since the devices are categorized in different groups, their energy use can be considered independent of each other. Equation (1) can thus be re-written as:

$$\sum_{a=1}^{|A|} \min \sum_{t=1}^{N_T} p_t \times P_{t,a} \times \Delta t. \tag{2}$$

The behavior of the buffer appliances can be formulated as:

$$\min \sum_{t}^{N_T} p_t \times P_{t,a} \times \Delta t \ \ \forall a \in A^{bf} \tag{3}$$

subject to,

$$\sum_{k+x_a}^{t=k} P_{t,a} = \sum_{k+x_a}^{t=k} P_{t,a}^0 \ \ \forall a \in A^{bf}, k \in T \tag{4}$$

$$\min_T P_{t,a}^0 < P_{t,a} < \max_T P_{t,a}^0 \ \ \forall a \in A^{bf}, t \in T \tag{5}$$

where x_a is the maximum buffer time for appliance a, $P_{t,a}^0$ is the original load of appliance a at time t, T is the set of the time steps in a day and A^{bf} is the set of the buffer appliances. The constraints in Equations (4) and (5) control that the total energy consumed by the device will remain constant and buffer time, x_a, could be altered depending on the price.

For time shifting devices, the behavior can be represented as the following optimization problem:

$$\min \sum_{t=1}^{N_T} p_t \times P_{t+\tau,a} \times \Delta t \quad \forall a \in A^{ts} \tag{6}$$

subject to,

$$\sum_{t=1}^{N_T} P_{t,a} = \sum_{t=1}^{N_T} P_{t+\tau,a} \quad \forall a \in A^{ts} \tag{7}$$

$$\tau_{min,a} < \tau < \tau_{max,a} \quad \forall a \in A^{ts} \tag{8}$$

where τ is the time shift of the appliance, $\tau_{min,a}$ and $\tau_{max,a}$ are the limits of the maximum allowable time shift and A^{ts} is the set of all time shifting appliances. The optimization problem of the time shifting appliances needs to use the time shift τ as the decision variable rather than the power at each time instant.

In this work, PV systems have been considered as curtailable only, and their behavior has been considered inelastic to dynamic price signals. More information of the device level modeling can be found in [6].

3.2. Price Adjustments

Generally, day-ahead dynamic price for an end-user consists of three distinct parts: the day-ahead market price (p_t^D), a flat network tariff (p^{DSO}) and taxes (p^{tax}). Thus, the day-ahead price at any given time, p_t can be expressed as,

$$p_t = p_t^D + p^{DSO} + p^{tax}. \tag{9}$$

In the Netherlands, the day-ahead market price for a particular day has an hourly resolution and is available at 12:00 on the day before after the closure of the day-ahead market. The network tariff is usually fixed and is around EUR 150 per year. On top of these, the energy tax is applied based on the volume of energy consumption [5].

In this work, we aim to vary the flat network tariff to influence the end-users for shifting the flexible appliances from the peak hours to off-peak moments.

3.2.1. Overloading Cost of the Transformer

Overloading of a transformer occurs due to higher loads at connection points such as, for example, with charging a large number of EVs or the operation of domestic HPs. Thermal overloading generally affects the insulation of the transformer windings; however, the involved dynamics in the loading enable the transformer to be overloaded for some time. Consequently, the amount of flexibility that needs to be procured should be aligned with the thermal status and the respective cost of overloading of the transformer.

Based on the provided load-profiles and historical values of the transformer load, the TA generates a set of probable loading scenarios to tackle the inherent uncertainties. For each of these scenarios, it estimates the imminent loss-of-life and respective overloading cost of the transformers. To do so, the following steps are followed:

Step 1: In this step, the TA generates the set of scenarios for probable loading. This is to counter the uncertainties associated with real-time loading, deviations from day-ahead schedule and forecast errors of the local generation technologies. In reality, correlations exist between the loads of consecutive hours. Therefore, instead of calculating single values for each time step, pseudo-random profiles of loading are generated based on Copula theory [33].

A copula is a multivariate probability distribution where the marginal-distributions are uniform. According to the Sklar's theorem, if those marginal distributions are continuous, the copula is unique. Let F be a 24-dimensional distribution function with continuous margins, $(F_1, F_2 \ldots \ldots, F_{24})$. A copula can be fitted, using the cumulative distribution functions (CDFs) of the historical load and the expected load profile provided by the aggregator. The marginal distribution here is the cumulative distribution functions generated by a Kernel Density Estimator (KDE). Then, the CDFs are used as inputs to generate the copula reflecting the correlation between the loads at different time steps. Next, the synthetic profiles generated from the fitted copula are transformed back to the original scale by applying the inverse cumulative distribution function. The whole process of scenario generation can be summarized by Figure 3.

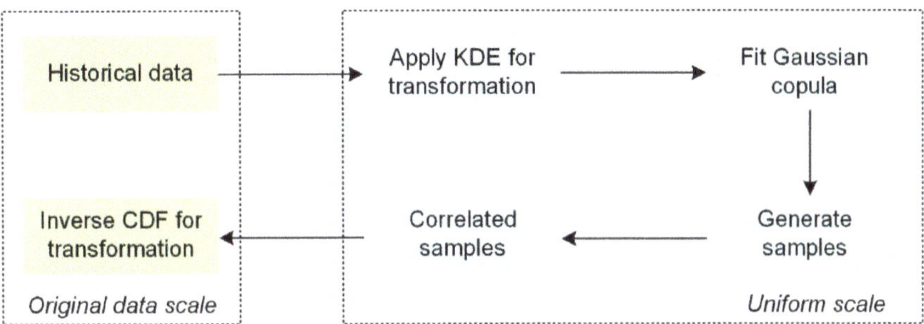

Figure 3. Generating day-ahead loading scenarios using copula.

Step 2: The expected load is converted to the resulting hottest-spot temperature, θ_H, of the transformer. According to IEEE Std C57.12.00-1993, the hottest-spot temperature is defined as the highest temperature of the winding at the operating condition, and is the main element for calculating the expected life of a transformer [34]. To determine this, a ratio called the load multiplex, K, is calculated and then used to calculate the top oil temperature rise, $\Delta\theta_{TO}$, and the hottest-spot temperature rise, $\Delta\theta_H$, as:

$$K = \frac{\text{Expected load}}{\text{Rated load}} \tag{10}$$

$$\Delta\theta_{TO} = \Delta\theta_{TO,R}\left[\frac{K^2 R + 1}{R + 1}\right] \tag{11}$$

$$\Delta\theta_H = \Delta\theta_{H,R}K^2 \tag{12}$$

where $\Delta\theta_{TO,R}$ is the top oil temperature rise at rated load, $\Delta\theta_{H,R}$ is hottest-spot temperature at the rated load and R is the ratio of load loss at rated load to no-load loss at rated load. Then, θ_H is found by summing the ambient temperature θ_A with the above mentioned temperature rises.

$$\theta_H = \theta_A + \Delta\theta_{TO} + \Delta\theta_H \tag{13}$$

Step 3: In this step, the ageing acceleration factor (F_{AA}) and equivalent ageing factor (F_{eqv}) are calculated for the particular combination of loads and temperature for a duration of thirty min. This is due to the fact that degradation of the insulation is realized when the transformer is generally overloaded for half-an hour.

$$F_{AA} = e^{\left[\frac{15000}{383} - \frac{15000}{\theta_H + 273}\right]} \tag{14}$$

$$F_{eqv} = \frac{\sum_{n=1}^{N} F_{AA}\Delta t_n}{\sum_{n=1}^{N} \Delta t_n} \tag{15}$$

where N is the number of time intervals, the duration of each is Δt_n hours and normal insulation life of the transformer is T_{inl} hours. According to IEEE standards, the normal insulation life of a well dried, oxygen free distribution transformer is 180,000 h or 20.55 years [2,16].

Step 4: Based on the calculated equivalent factor, per unit loss of life of the transformer T_{lol} is determined.

$$T_{lol} = \frac{F_{eqv}t}{T_{inl}}$$ (16)

The aging cost C_{ag} can be determined by the loss of life with the total owning cost (TOC) C_o of the transformer. The TOC method is considered to be one of the most cost and resource efficient methods for economic analysis of a transformer [35]. In addition to the initial cost of the transformer, TOC considers the operation and maintenance cost of the transformer and is calculated over the life span of the asset. The TOC can be determined from purchase cost C_P, cost of no-load loss C_{NL} and cost of load loss C_{LL} of the transformer [35].

$$C_o = C_P + C_{NL} + C_{LL}$$ (17)

$$C_{ag} = T_{lol}C_o$$ (18)

If C_{ag} is greater than the aging cost at nominal rating of the transformer $C_{ag,R}$, then the overloading cost C_{OL} is determined from the arithmetic difference. Otherwise, the overloading cost is assumed to be zero.

$$C^{OL} = \begin{cases} C_{ag} - C_{ag,R} & \text{when } C_{ag} > C_{ag,R} \\ 0 & \text{otherwise} \end{cases}$$ (19)

3.2.2. Calculation of Dynamic Tariff

The TA calculates the dynamic tariff that satisfies the thermal constraints for each of the loading scenarios $s \in S$. This can be mathematically presented as the following optimization problem:

$$\min \sum_{s=1, s \in S}^{s=s_N} \sum_{t=1, t \in T}^{t=t_N} (p_t^{DSO} - p^{DSO})P_t^s \ \forall s \in S \ \text{ if } (C_{t,s}^{OL} > 0)$$ (20)

subject to,

$$\overline{p_t^{DSO}} = \overline{p^{DSO}}$$ (21)

$$p_t^{DSO} = \frac{1}{5} \sum_{t_f=t-2}^{t+2} p_t^{DSO} + p_{adj} \text{ if } C_{t_f,s}^{OL} > 0$$ (22)

$$p_{adj,t} \leq \max\left(C_{t,s}^{OL}\right).$$ (23)

p_t^{DSO} is the decision variable and denotes the dynamic network tariff at each time step. P_t^s is the load at time $t \in T$ in scenario, s. Constraints in Equation (21) dictate that the average network tariff should be the same for both normal operations and tariff-based DR cases. Constraints in Equations (22) and (23) limit the range of tariff adjustment in time and monetary values, respectively.

The resulting dynamic network tariff is sent to the Aggregator for updating the price for the end-users. This process is continued as an iterative process until the issues are resolved. In this work, we have limited the process to up to ten iterations. The whole process can be represented by the flowchart as shown in Figure 4.

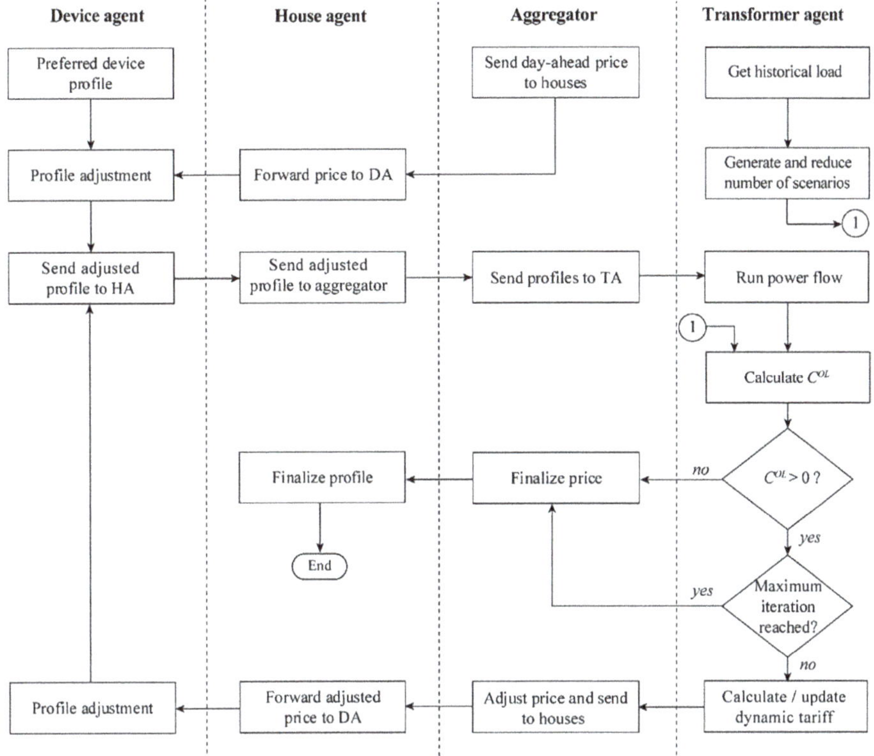

Figure 4. Flowchart depicting the processes.

4. Simulation Setup

4.1. Load Modeling

A bottom-up approach is adopted for the simulation test case. The residential loads are classified into seven categories: base load, EV, PV, heat pump, washing machine, dishwasher and refrigerator. The base load is considered to be comprised of the non-controllable devices. The required EV charge is modeled based on the driving distance, arrival and departure time [36]. The charging rate is assumed to range between 3 and 8 kW. For the PV, the rated power varies between 2 and 5.5 kW per household. The heat pumps have a rated power of 0.7 kW to 2 kW, with an additional 2 kW resistive heating element. The washing machine, dishwasher and refrigerator have a rated power in the range of 0.6–2 kW, 0.5–1.11 kW and 0.035–0.140 kW, respectively. For more information about the loads, the reader is referred to [6]. A flat network tariff of EUR 0.06 is considered for p^{DSO} [37]. The network tariff is adjusted in such a way so that the daily average is kept fixed at EUR 0.06.

4.2. Network

A modified version of the IEEE European LV test network is used for the case study [38]. The network hosts 55 households with single phase connections. A 250 kVA, 11 kV/0.416 kV transformer supplies the LV segment from the MV bus. The resistance and reactance of the windings are 0.4% and 4% of the base values at the MV side, respectively. An additional LV feeder with an aggregated peak load of 100 kW is assumed at the substation. A constant power factor of 0.95 is used for the power flow calculation. Per unit values in the results are calculated considering the transformer rating as the base value.

4.3. Simulation Platform

The simulations have been performed with a time step of 15 min. The MAS-based coordination is done in a MATLAB environment. Due to the relatively simpler form of the local optimization problems, the MATLAB Optimization Toolbox has been used to solve them. The power flow calculations are performed using the EPRI distribution system simulator, OpenDSS [39].

5. Numerical Results

In this section, simulation results are presented. First, the generated scenarios are briefly discussed, followed by a discussion on the effect of dynamic price on residential demand. Next, the impacts of the dynamic network tariff are explained, followed by the monthly performance of the proposed approach in terms of overall efficiency of managing the congestion.

5.1. Scenarios of Loading

Figure 5 shows the scenarios generated based on the historical loading data of the transformer and the scheduled demand for the following day based on the day-ahead dynamic price, as supplied by the aggregator. First, a set of 1000 scenarios are generated using Gaussian copula to represent the correlation among different time steps based on the data of previous one month. In order to simplify the calculation, this set of scenarios is reduced to 20 that closely represent the behavior of all the members of the set. Figure 5c shows that the set of reduced scenarios can capture the dynamics of all the time steps of the following day based on the day-ahead price. Thus, the set of reduced scenarios can address the uncertainties with the loading of the following day.

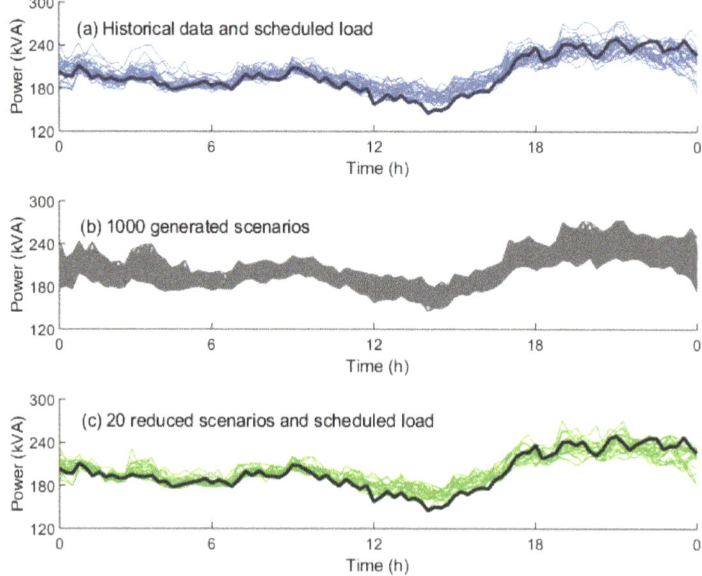

Figure 5. Transformer loading, (**a**) historical values of one month previous and based on scheduled loads for the following day, (**b**) generated scenarios based on Gaussian copula to represent the correlation and (**c**) reduced scenarios and the scheduled load.

5.2. Effects of Dynamic Market Price

Figure 6 illustrates the effects of the dynamic market price on the network load as a whole. The initial load profile represents the projected profiles of the households. The HAs adapt the initial profile based on the identical price from the aggregator. The price-adjusted profile shows distinct peaks when the price levels are lower. The HAs essentially minimize the total energy cost by optimizing the daily usage of all the appliances. To this end, the flexible appliances are mostly scheduled at times of lower price. Consequently, a reduced domestic energy cost is realized by the dynamic price levels at the expense of higher overall peak load.

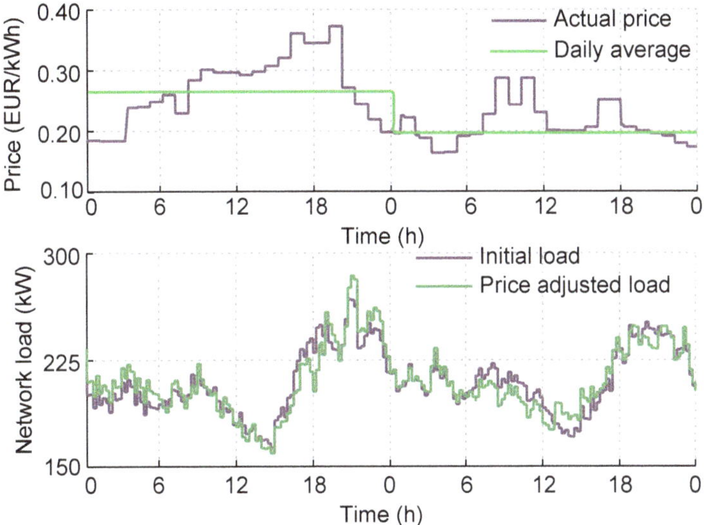

Figure 6. Effects of dynamic price on day-ahead scheduled load.

5.3. Dynamic Network Tariff for Peak Reduction

In order to resolve the congestion, the TA requests flexibility from the aggregator. This is done by introducing the variable network tariff, as discussed in Section 3.2. As the coordinator of the demand flexibility, the aggregator adjusts the price sent to the HAs by adding the variable tariff instead of the flat counterpart. Flexible appliances of the end-users react to this newly available price and try to shift loads from the peak moments. Figures 7 and 8 show the change in price and the resulting adjusted load profiles for two consecutive days, respectively. To make use of the decreasing price at the evening of the first day, EVs start charging almost simultaneously and, as a result, they cause congestion in the transformer. Similarly, the coincidence of the HP operations with the conventional evening peak leads to congestion on the second day.

Based on the overloading cost and expected loading scenarios, the TA changes the network tariff. The off-peak price level is further reduced due to lower tariff in order to maintain a constant daily average. The HAs respond to the new price levels by re-optimizing the flexible loads. The effect of the increase is reflected in the loads, as the peak load on the first day is lowered to 0.98 p.u. from 1.04 p.u. In both cases, the total energy consumption remains 4.95 MWh, since the flexible loads are merely shifted in time.

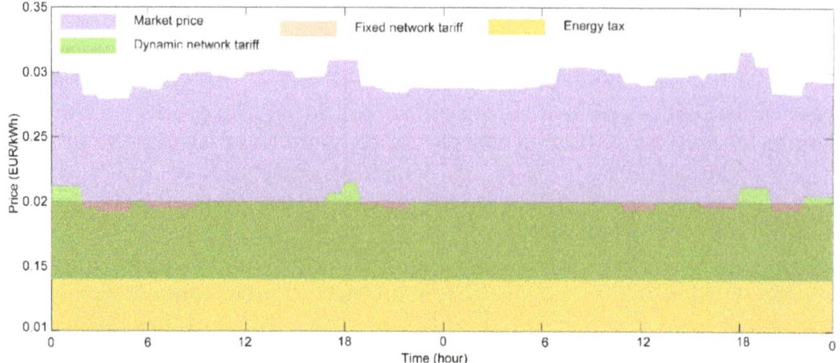

Figure 7. Electricity price breakdown in different cases.

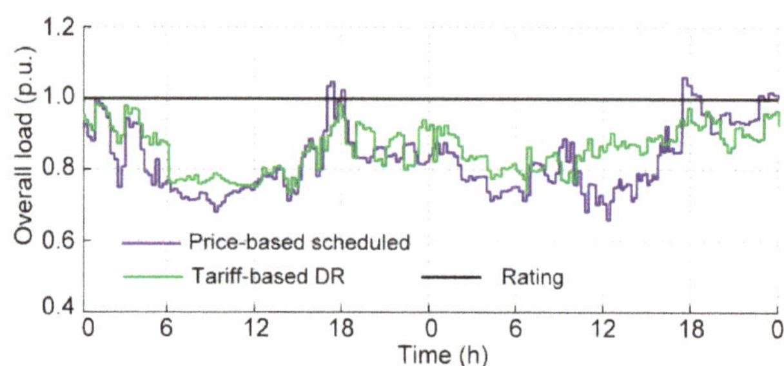

Figure 8. Simulated network load for two consecutive days.

5.4. Monthly Assessment

The overall monthly performance of the proposed approach is analyzed in terms of the congestion duration, network peak load and total supplied energy. This ensures the efficiency of the approach for resolving the congestion while maintaining the quality of supply. Table 1 depicts the comparative values of the uncontrolled load, the price-based scheduling and the proposed tariff-based DR scheme. The congestion duration is considerably reduced with the proposed approach at the expense of a drop in supplied energy. This happens due to the fact that some of the flexible appliances shift their energy consumption at later hours (possibly the next day) and, hence, were not considered within the simulation time window. The average monthly energy cost of the end-users has been reduced for the proposed method, although the objective of the proposed approach is to tackle imminent congestion in the network, and not to reduce the energy cost of the end-users. In reality, the end-users will not have to face the effect, since the daily average of the tariff will be the same. In other words, they will keep paying the same amount for the network tariff as before. The change in the cost is generated from the differences in day-ahead price at corresponding time steps.

Table 1. Comparative monthly assessment among different schemes.

Performance Indicators	Uncontrolled Load	Price-Based Scheduling	Tariff-Based Dr
Maximum load (p.u.)	1.07	1.07	1.05
Duration of congestion (h)	11.50	9.75	1.75
Supplied energy (MWh)	148.22	148.34	148.07
Average domestic cost (EUR)	43.88	41.23	42.02

The effects are also evident in the monthly load duration curve, as shown in Figure 9. The controlled case demonstrates a notable reduction in the peak demand and shifts the loads when the network tariff is lower. Since the total energy consumption is of similar order, the comfort levels of the end-users were also ensured.

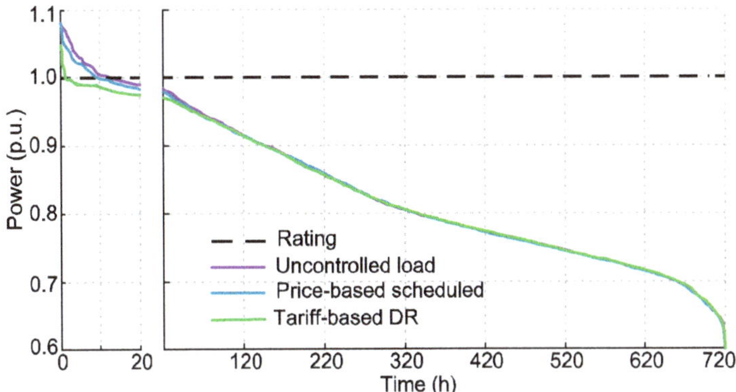

Figure 9. Monthly load duration curve.

6. Conclusions

A dynamic network tariff has been introduced in this paper that utilizes a hierarchical MAS-based architecture with distributed intelligence. The price-responsive behaviors of the loads have been modeled using a detailed bottom-up Markov Chain Monte Carlo approach. The loads are scheduled based on the dynamic prices available from the aggregator. If a case of network congestion is expected, the DSO adjusts the network tariff part of the price by varying it throughout the day. This is done on the basis of an estimated overloading cost resulting from the probable loading scenarios. The loading scenarios are generated using the Gaussian copula of the expected load from the aggregator and historical loading values of one month previous. The network tariff is raised around the peak hours while keeping the daily average fixed.

The proposed approach has been evaluated through simulations for 55 households in a modified form of the IEEE European LV test feeder. Simulation results reveal a notable relation between the dynamic price and congestion, which has been reported by a number of previous studies. The proposed approach appears to be efficient at managing congestions, as the total monthly congestion duration has been reduced up to 82%. The monthly simulation has been performed on an Intel Core i7 computer with 8 GB of RAM. The proposed approach requires a simulation time of approximately 15 min compared to 12 min for the uncontrolled case.

Future research on this topic will be directed to efficient methods for local voltage control. The functionalities of the aggregator will be further updated with smarter learning techniques for more accurate price adjustments.

Author Contributions: Literature Research, N.H. and A.T.; Conceptualization, N.H.; Methodology, N.H.; Visualization, N.H. and A.T.; Writing original draft, N.H. and A.T.; Writing review and editing, N.H., P.N. and G.P. All authors have read and agreed to the published version of the manuscript.

Funding: This work is part of the research program ERA-net Smart Energy Systems with project number 651.001.012, which is financed by the Netherlands Organization for Scientific Research (NWO).

Conflicts of Interest: The authors declare no conflict of interest.

Nomenclature

List of Abbreviations

CDF	Cumulative Distribution Function		KDE	Kernel Density Estimator
DA	Device Agent		LV	Low-voltage
DR	Demand Response		MAS	Multi-Agent Systems
DS	Dynamic Subsidy		MV	Medium-voltage
DSO	Distribution System Operator		PV	Photovoltaics
EPRI	Electric Power Research Institute		RES	Renewable Energy Sources
EV	Electric Vehicle		TA	Transformer Agent
HA	House Agent		TOC	Total Owning Cost
HP	Heat Pump			

List of Symbols

Indices

a	index of appliances		θ_H	hottest spot temperature
t	index of time steps		θ_A	ambient temperature
bf	index of buffer appliances		F_{AA}	ageing acceleration factor
ts	index of time shifting appliances		F_{eqv}	equivalent factor
D	index of day		C_{ag}	ageing cost
s	index of scenarios		T_{lol}	loss of life
			C^{OL}	overloading cost

Sets — **Parameters**

A	set of appliances		N_t	number of daily time steps
T	set of all time steps		Δt	duration of a time step in hour
S	set of all scenarios		x	buffer time

Variables

p	electricity price		τ_{min}	lower limit of allowable time shift
P	active power		τ_{max}	upper limit of allowable time shift
k	time shift of buffer appliances		$\Delta\theta_{TO,R}$	top oil temperature rise at rated load
τ	time shift of time shifting appliances		$\Delta\theta_{H,R}$	hottest-spot temperature at rated load
K	load multiplex		R	ratio of load loss to no-load loss
$\Delta\theta_{TO}$	top oil temperature rise		T_{inl}	normal insulation life of transformer
$\Delta\theta_H$	hottest spot temperature rise		C_P	purchase cost
			C_{NL}, C_{LL}	cost of no-load loss and load loss

References

1. Haque, A.N.M.M.; Nguyen, P.H.; Vo, T.H.; Bliek, F.W. Agent-based unified approach for thermal and voltage constraint management in LV distribution network. *Electr. Power Syst. Res.* **2017**, *143*, 462–473. [CrossRef]
2. Haque, A.N.M.M.; Shafiullah, D.S.; Nguyen, P.H.; Bliek, F.W. Real-Time Congestion Management in Active Distribution Network based on Dynamic Thermal Overloading Cost. In Proceedings of the Power Systems Computation Conference (PSCC), Genoa, Italy, 20–24 June 2016; pp. 1–7.
3. Veldman, E.; Verzijlbergh, R.A. Distribution Grid Impacts of Smart Electric Vehicle Charging from Different Perspectives. *IEEE Trans. Smart Grid* **2015**, *6*, 333–342. [CrossRef]
4. Haque, A.N.M.M. *Smart Congestion Management in Active Distribution Networks*; Eindhoven University of Technology: Eindhoven, The Netherlands, 2017.
5. Klaassen, E.A.M. *Demand Response Benefits from a Power System Perspective*; Eindhoven University of Technology: Eindhoven, The Netherlands, 2016.
6. Haque, A.; Nijhuis, M.; Ye, G.; Nguyen, P.; Bliek, F.; Slootweg, J. Integrating Direct and Indirect Load Control for Congestion Management in LV Networks. *IEEE Trans. Smart Grid* **2017**, *10*, 741–751. [CrossRef]
7. Li, Y.; Li, N. Mechanism design for reliability in demand response with uncertainty. In Proceedings of the American Control Conference, Seattle, WA, USA, 24–26 May 2017.
8. Ma, H.; Parkes, D.C.; Robu, V. Generalizing demand response through reward bidding. In Proceedings of the International Joint Conference on Autonomous Agents and Multiagent Systems AAMAS, São Paulo, Brazil, 8–12 May 2017.

9. Hu, J.; You, S.; Lind, M.; Østergaard, J. Coordinated charging of electric vehicles for congestion prevention in the distribution grid. *IEEE Trans. Smart Grid* **2014**, *5*, 703–711. [CrossRef]

10. Liu, W.; Wu, Q.; Wen, F.; Ostergaard, J. Day-ahead congestion management in distribution systems through household demand response and distribution congestion prices. *IEEE Trans. Smart Grid* **2014**, *5*, 2739–2747. [CrossRef]

11. Nguyen, P.H.; Kling, W.L.; Ribeiro, P.F. A game theory strategy to integrate distributed agent-based functions in smart grids. *IEEE Trans. Smart Grid* **2013**, *4*, 568–576. [CrossRef]

12. Palensky, P.; Dietrich, D. Demand side management: Demand response, intelligent energy systems, and smart loads. *IEEE Trans. Ind. Inf.* **2011**, *7*, 381–388. [CrossRef]

13. Ghazvini, M.A.F.; Lipari, G.; Pau, M.; Ponci, F.; Monti, A.; Soares, J.; Castro, R.; Vale, Z. Congestion management in active distribution networks through demand response implementation. *Sustain. Energy Grids Netw.* **2019**, *17*, 100185. [CrossRef]

14. Garcia, T.S.; Shafie-Khah, M.; Osório, G.J.; Catalão, J.P.S. Optimal bidding strategy of responsive demands in a new decentralized market-based scheme. In Proceedings of the 17th IEEE International Conference on Environment and Electrical Engineering and 1st IEEE Industrial and Commercial Power Systems Europe, Milan, Italy, 6–9 June 2017.

15. Chai, B.; Chen, J.; Yang, Z.; Zhang, Y. Demand response management with multiple utility companies: A two-level game approach. *IEEE Trans. Smart Grid* **2014**, *5*, 722–731. [CrossRef]

16. Haque, A.N.M.M.; Nguyen, P.H.; Bliek, F.W.; Slootweg, J.G. Demand response for real-time congestion management incorporating dynamic thermal overloading cost. *Sustain. Energy Grids Netw.* **2017**, *10*, 65–74. [CrossRef]

17. Nizami, M.S.H.; Haque, A.N.M.M.; Nguyen, P.H.; Bliek, F.W. HEMS as Network Support Tool: Facilitating Network Operator in Congestion Management and Overvoltage Mitigation. In Proceedings of the 16th International Conference on Environment and Electrical Engineering (EEEIC), Florence, Italy, 6–8 June 2016; pp. 1–6.

18. Torbaghan, S.S.; Gibescu, M.; Rawn, B.G.; van der Meijden, M. A market-based transmission planning for HVDC grid—Case study of the North Sea. *IEEE Trans. Power Syst.* **2015**, *30*, 784–794. [CrossRef]

19. Hu, J.; Saleem, A.; You, S.; Nordström, L.; Lind, M.; Østergaard, J. A multi-agent system for distribution grid congestion management with electric vehicles. *Eng. Appl. Artif. Intell.* **2015**, *38*, 45–58. [CrossRef]

20. Zhang, C.; Ding, Y.; Østergaard, J.; Bindner, H.W.; Nordentoft, N.C.; Hansen, L.H.; Brath, P.; Cajar, P.D. A flex-market design for flexibility services through DERs. In Proceedings of the 4th IEEE/PES Innovative Smart Grid Technologies Europe, Copenhagen, Denmark, 6–9 October 2013.

21. Huang, S.; Wu, Q.; Cheng, L.; Liu, Z.; Zhao, H. Uncertainty Management of Dynamic Tariff Method for Congestion Management in Distribution Networks. *IEEE Trans. Power Syst.* **2016**, *3053*, 1–12. [CrossRef]

22. Huang, S.; Wu, Q. Dynamic Subsidy Method for Congestion Management in Distribution Networks. *IEEE Trans. Smart Grid* **2018**, *9*, 2140–2151. [CrossRef]

23. Zhao, J.; Wang, Y.; Song, G.; Li, P.; Wang, C.; Wu, J. Congestion Management Method of Low-Voltage Active Distribution Networks Based on Distribution Locational Marginal Price. *IEEE Access* **2019**, *7*, 32240–32255. [CrossRef]

24. Lampropoulos, I.; van den Broek, M.; van der Hoofd, E.; Hommes, K.; van Sark, W. A system perspective to the deployment of flexibility through aggregator companies in the Netherlands. *Energy Policy* **2018**, *118*, 534–551. [CrossRef]

25. Biggar, D.; Reeves, A. Network Pricing for the Prosumer Future: Demand-Based Tariffs or Locational Marginal Pricing? In *Future of Utilities–Utilities of the Future: How Technological Innovations in Distributed Energy Resources Will Reshape the Electric Power Sector*; Academic Press: Cambridge, MA, USA, 2016.

26. Torbaghan, S.S.; Blaauwbroek, N.; Nguyen, P.; Gibescu, M. Local market framework for exploiting flexibility from the end users. In Proceedings of the 2016 13th International Conference on the European Energy Market (EEM), Porto, Portugal, 6–9 June 2016; pp. 1–6. [CrossRef]

27. Eid, C.; Codani, P.; Perez, Y.; Reneses, J.; Hakvoort, R. Managing electric flexibility from Distributed Energy Resources: A review for incentives, aggregation and market design. *Renew. Sustain. Energy Rev.* **2015**, *64*, 237–247. [CrossRef]

28. Nwulu, N.I.; Xia, X. Optimal dispatch for a microgrid incorporating renewables and demand response. *Renew. Energy* **2017**, *101*, 16–28. [CrossRef]

29. Pratt, B.A.; Krishnamurthy, D.; Ruth, M. Transactive Home Energy Management Systems. *IEEE Electrif. Mag.* **2016**, *4*, 8–14. [CrossRef]

30. Minniti, S.; Haque, N.; Nguyen, P.; Pemen, G. Local markets for flexibility trading: Key stages and enablers. *Energies* **2018**, *11*, 3074. [CrossRef]

31. Nijhuis, M.; Gibescu, M.; Cobben, J.F.G. Bottom-up Markov Chain Monte Carlo approach for scenario based residential load modelling with publicly available data. *Energy Build.* **2016**, *112*, 121–129. [CrossRef]

32. Flexible Power Alliance Interface (FPAI). Available online: http://www.flexiblepower.org/downloads/ (accessed on 21 December 2019).

33. Ni, F.; Nguyen, P.H.; Cobben, J.F.G. Basis-Adaptive Sparse Polynomial Chaos Expansion for Probabilistic Power Flow. *IEEE Trans. Power Syst.* **2017**, *32*, 694–704. [CrossRef]

34. ABB. *Distribution Transformer Handbook IEC/CENELEC Related Specifications ABB*; ABB: Zurich, Switzerland, 2003.

35. Downing, D.J.; McConnell, B.W.; Barnes, P.R.; Hadley, S.W.; van Dyke, J.W. *Economic Analysis of Efficient Distribution Transformer Trends*; Oak Ridge National Laboratory: Oak Ridge, TN, USA, 1998.

36. Verzijlbergh, R.A.; Lukszo, Z.; Veldman, E.; Slootweg, J.G.; Ilic, M. Deriving electric vehicle charge profiles from driving statistics. In Proceedings of the IEEE Power & Energy Society General Meeting, Detroit, MI, USA, 24–28 July 2011; pp. 1–6.

37. Verzijlbergh, R.A. *The Power of Electric Vehicles*; Delft University of Technology: Delft, The Netherlands, 2013.

38. IEEE PES Distribution Test Feeder Working Group. Distribution Test Feeders. Available online: https://site.ieee.org/pes-testfeeders/resources/ (accessed on 21 December 2019).

39. EPRI. OpenDSS, Distribution System Simulator. Available online: http://sourceforge.net/projects/electricdss (accessed on 26 June 2018).

Article

An Optimal Energy Management System for Real-Time Operation of Multiagent-Based Microgrids Using a T-Cell Algorithm

Fatima Zahra Harmouch [1,2], Ahmed F. Ebrahim [2], Mohammad Mahmoudian Esfahani [2], Nissrine Krami [1], Nabil Hmina [1] and Osama A. Mohammed [2,*]

[1]   Laboratory Systems Engineering (LGS) ENSA, Ibn Tofail University, Kenitra, B.P 241, Morocco
[2]   Energy Systems Research Laboratory, Department of Electrical and Computer Engineering, Florida International University, Miami, FL 33174, USA
*   Correspondence: mohammed@fiu.edu; Tel.: +1-1305-348-3040

Received: 25 June 2019; Accepted: 30 July 2019; Published: 3 August 2019

Abstract: The real-time operation of the energy management system (RT-EMS) is one of the vital functions of Microgrids (MG). In this context, the reliability and smooth operation should be maintained in real time regardless of load and generation variations and without losing the optimum operation cost. This paper presents a design and implementation of a RT-EMS based on Multiagent system (MAS) and the fast converging T-Cell algorithm to minimize the MG operational cost and maximize the real-time response in grid-connected MG. The RT-EMS has the main function to ensure the energy dispatch between the distributed generation (DG) units that consist in this work on a wind generator, solar energy, energy storage units, controllable loads and the main grid. A modular multi-agent platform is proposed to implement the RT-EMS. The MAS has features such as peer-to-peer communication capability, a fault-tolerance structure, and high flexibility, which make it convenient for MG context. Each component of the MG has its own managing agent. While, the MG optimizer (MGO) is the agent responsible for running the optimization and ensuring the seamless operation of the MG in real time, the MG supervisor (MGS) is the agent that intercepts sudden high load variations and computes the new optimum operating point. In addition, the proposed RT-EMS develops an integration of the MAS platform with the Data Distribution Service (DDS) as a middleware to communicate with the physical units. In this work, the proposed algorithm minimizes the cost function of the MG as well as maximizes the use of renewable energy generation; Then, it assigns the power reference to each DG of the MG. The total time delay of the optimization and the communication between the EMS components were reduced. To verify the performance of our proposed system, an experimental validation in a MG testbed were conducted. Results show the reliability and the effectiveness of the proposed multiagent based RT-EMS. Various scenarios were tested such as normal operation as well as sudden load variation. The optimum values were obtained faster in terms of computation time as compared to existing techniques. The latency from the proposed system was 43% faster than other heuristic or deterministic methods in the literature. This significant improvement makes this proposed system more competitive for RT applications.

Keywords: smart microgrid; energy management system; multi-agent system; real-time optimization; immune system algorithm; economic dispatch

---

1. Introduction

Technology has shaped our modern life in many aspects: starting by smart TV, passing by smart phones and ending by Smart Grid. The power system which is conventionally structured as generation, transmission and distribution is reshaped as an agglomeration of Microgrids (MGs). The idea behind such a new structure is the increasing distributed energy resource (DER) mainly renewable energy that has been installed in the low voltage (LV) networks. Injecting power into the distribution power system is a new phenomenon which must be dealt with. This is where the smart cyber systems interfere. Information technologies has been integrated with the power system enabling, as consequence, the smart control of DG and make it possible for the MG to operable autonomously without being connected to the grid. Besides, such structure favorizes the interaction of the costumer to become an active player in the system and control its load depending on the distribution network operator.

The information system responsible for the control and economical optimization of MG is known as the energy management system (EMS). The EMS comprises three parts. The primary, secondary and tertiary control. More details on the primary, secondary could be found in [1,2]. This work focuses on the tertiary control. One of the most crucial tasks of a real-time EMS in grid-connected MG is to satisfy the minimum total operation cost while meeting the energy balance between different DER, storage units, controllable loads and the power exchanged with the main grid under their specific constraints. This is called in the literature review as the economic dispatch problem or the economic load dispatch problem. To solve this issue, several optimization methods have been used in the past decades. All of them can be categorized into two main categories: Classical and evolutionary algorithms.

In the conventional power system, the problem is modeled as a quadratic cost function which is solved by classical methods such as Lagrange multiplier method, base point participation factor, lambda iteration, Newton's method, gradient method, mixed integer linear programming, and linear programming [3]. Most of previous works presented a power dispatch of thermal generators which makes the cost function considered not suitable for MG optimization [4,5]. Besides, some of those used technics suffer from some limitations such as the assumption that the incremental cost curves of the generation units are monotonically increasing piecewise linear functions and the high dependency on the specific mathematical model. Even if dynamic programming could be used to solve the economic dispatch, it presents a complex solutions that is not appropriate for real time application [6]. Furthermore, it is not possible to apply such methods in real time specially with a MG containing a large number of components.

In this respect, evolutionary algorithms were introduced to overcome those limitations. The heuristic evolutionary algorithms are inspired from the natural and social behavior of animals and organisms [3]. For instance, genetic algorithm, improved genetic algorithm were based on genes behavior, while evolutionary programming, ant colony metaheuristic, artificial bee colony, particle swarm optimization and artificial neural network are based on swarm behavior of insects and micro-organisms. However, even if some of these methods present competitive results, still some of them suffer from the curse of local optimality in some cases [7]. The literature review reveals that most of previous work were considering a day ahead optimization and not taking the real time constraint into consideration [8]. More recent works present the real time operation of energy management in MG [9,10]. However, some of the work were based on simulations and did not experiment the RT operation in real testbed, while other work did not consider the point of single failure of the system. A RT framework for EMS of MG using MAS were considered in [11], however there was no cost optimization in the system.

Since deterministic optimization methods are time consuming to reach the optimal solution for problems with large dimensions, evolutionary algorithms were introduced to overcome this limitation. The main advantage of these solutions is that they solve complex problems with non-deterministic polynomial problem. Furthermore, in the MG context, while the EMS is running in real time, solving the ED optimization time is also a crucial parameter. Thus, looking for a fast optimization technique is essential.

In this work, the T-Cell algorithm is presented as a heuristic method inspired by immunobiology to solve ED in RT. The algorithm was first introduced in [12]. The T-Cell algorithm is inspired by the mediated immune cells in the human body. The T-Cell or as known in the immunobiology, helper

T-Cells play a key role in the adaptative immune response. During their life, helper T-cells experience three phases. During the first phase of their life, they develop in the Thymus, then they immigrate and start circulating between the bloodstream and lymphoid tissue. At this stage, they are known as a naïve helper T-cells. Then in the second phase, the naïve T-cells encounter an antigen-presenting cell. This latter binds to the peptide and protein receptors at the cell surface which results in the T-Cell activation. Once activated, the cell starts proliferating and differentiating. This is the third phase. Helper T-Cell differentiate into either T helper 1 or T helper 2 effector cells, depending on their environment's composition. Each type of effector cell helps to eliminate the antigen presented in the activation process [13]. Therefore, during an antibody intrusion, the T-Cells proliferate by generating other clones of themselves; then, each clone differentiates by acquiring new proprieties to destroy the intrusion [14]. The authors in [15] tested the performance of T-Cell algorithm compared to other optimization heuristics. The algorithm presents a competitive result in term of convergence and execution time.

However, the optimization is just one part of an EMS. The EMS in the context of MGs deals with all management aspects of MG. It must coordinate between different DERs, supervise the network in case of load or generation fluctuation and make sure that energy balance is satisfied all time. A multi-agent system (MAS) is well suited in this context. MAS knew lot of improvement during last years. The nature of the modular and decentralized structure increase researchers' interest to apply them in power system context [16]. An agent is nothing but a cyber or cyber physical entity that can act autonomously according to its environment, take decisions and achieve its goals and interacts with its peers [17].

The distributed nature of MAS makes it suitable for MG environment. Furthermore, their flexibility, extensibility, and fault tolerance have pertinently justified their adoption. Flexibility refers to the agent's ability to choose the most appropriate action; extensibility means the system can be extended easily; fault tolerance refers to the inherent redundancy mechanism build in MAS [18]. The central management of MG has been criticized for a long time by its drawback the single point of failure. However, using a multiagent platform that implements fault tolerance capabilities would resolve this issue. For these reasons, MAS was adopted in many areas of the power system. The classic bidding mechanism of electricity, for instance, used the MAS as an efficient negotiation tool [19–22]. Finally, MAS was also implemented as a fundamental MG and smart building management platform [10,23–26].

For these reasons, this work harvests the advantage of using heuristic algorithm that is inspired from the immune system, to optimize the ED and reaches the optimum in less time even in complex system with large dimensions. Besides, the use of MAS makes the system overcome the drawback of single point of failure and adds fault tolerance feature in addition to the ability of easy extensibility in case of future system growth.

1.1. Target

The present work's target is to implement the T-Cell algorithm as optimization methods for DED within MG based on agents' paradigm, and interoperability capabilities. The main contributions of this work are as follows:

- The implementation of an enhanced version of the T-Cell algorithm for RT-EMS within MG;
- The integration of the Multiagent EMS with the testbed lab using DDS;
- Fault tolerance implementation of the multiagent EMS;
- The experimental validation and implementation of this architecture into a real MG;
- Competitive time delay optimization for real-time application.

This work presents a new optimization algorithm in the MG context. Although, some heuristic optimizations were used in the EMS in MG, other system's features were not taken into consideration such as the central aspect of the EMS or the failure case management. Excluding the fault case scenario makes the system vulnerable and suffers from the single point of failure. Moreover, the authors objective is to design a comprehensive system that tackles several aspects of the MG EMS: Optimization and communication as a cyber physical system.

1.2. Materials

The present work uses different materials to implement the proposed system. The MAS was adopted for the EMS within the MG. The multiagent platform comprises three main agent's categories. Management agent, backup agent, and interoperability agents. The management agent comprises the MG optimizer (MGO) and the MG supervisor (MGS). The backup agent is the MGO backup agent. The communication agents correspond to MG unit agents that manage communication between the RT-EMS and the real components of the MG. Presentation of each agent will be explained in section four with more details.

The Data Distribution Service (DDS) middleware was also proposed as interoperability solution between the agent part and the hardware part is assured by the. The DDS is a middleware protocol and standard for data-centric communication for distributed real-time applications. It facilitates the development and integration of a distributed system and data exchange. The RT-EMS integrates the DDS components to communicate the computed references to the hardware part.

1.3. Method

The optimization of the DED in this work is solved by using the T-Cell heuristic algorithm inspired by the body immune system. This algorithm mimics the human body defense mechanism. The heuristic technics based on biological process has several advantages compared to classical deterministic methods. One of the main advantages of these solutions is that they solve complex problems with non-deterministic polynomial problem. The execution time of the optimization is improved and the same time an optimal solution is reached. These reasons are behind the motivation of using the T-Cell as techniques to solve our DED. Besides, the comparison of our algorithm to other existing ones shows better results in term of optimum computation and time delay which makes it competitive for RT applications.

Section 2 will cover the problem statement that includes the mathematical implementation of dynamic economic dispatch (DED). Section 3 describes the T-Cell algorithm that is implemented in the energy management system (EMS). Section 4 of this paper discusses the proposed multiagent platform that was designed in this study. Lastly, Sections 5 and 6 present and address the findings and concluding remarks on applicability, reliability, and time delay.

2. Problem Formulation

As mentioned before, the EMS is the information system responsible for the economical optimization and the steady operation of the MG. Its main function at a single MG is to ensure the dynamic economic dispatch (DED) between generators of the system. The DED problem consists of allocating the total demand among generating units so that the production cost is minimized [27]. However, it is important to stress that the DED in MG context is more straightforward than the DED in a conventional power system. Indeed, the MG has much fewer units than the whole system. Also, the DED in MG does not have a wide variety of conventional fuel generators and at the same time, tries to integrate more renewable energy RE resources. The hypotheses considered in this work are as follows:

- The distribution of power losses neglected;
- The electric water heater is considered a controllable load;
- The reactive power flow is neglected;
- The MG is operating in connected mode.

The objective function of the DED optimization problem is defined as:

$$\min \sum_{t=1}^{m} \left(F_t^{pv} + F_t^{w} + F_t^{ES-} + F_t^{G-} - F_t^{G+} - F_t^{ES+} - F_t^{l} - F_t^{cl} \right) \cdot \Delta t \tag{1}$$

m represents the optimization period. F_t^{pv} and F_t^{w} represent the cost function of the PV panels and the wind turbine respectively; F_t^{ES-} and F_t^{ES+} represent the cost function of the battery storage in discharging and charging mode respectively; F_t^{G-} and F_t^{G+} represent the cost function of the power imported from the grid and exported to the grid respectively; F_t^{l} and F_t^{cl} represent the cost function of the of non-controllable loads and controllable load of the EWH respectively. The generation cost function of each unit is determined by:

$$F_t^{pv} = \sum_{k=1}^{n_{pv}} \pi_t^{k,pv} P_t^{k,pv} \tag{2}$$

$$F_t^{w} = \sum_{k=1}^{n_{w}} \pi_t^{k,w} P_t^{k,w} \tag{3}$$

$$F_t^{ES-} = \sum_{k=1}^{n_{ES}} \pi_t^{k,ES-} \cdot (1 - \theta_{ES}) \cdot P_t^{k,ES-} \tag{4}$$

$$F_t^{ES+} = \sum_{k=1}^{n_{ES}} \pi_t^{k,ES+} \cdot \theta_{ES} \cdot P_t^{k,ES+} \tag{5}$$

$$F_t^{G-} = \pi_t^{G-} \cdot (1 - \theta_{G}) \cdot P_t^{G-} \tag{6}$$

$$F_t^{G+} = \pi_t^{G+} \cdot \theta_{G} \cdot P_t^{G+} \tag{7}$$

$$F_t^{l} = \sum_{k=1}^{n_{l}} \pi_t^{k,l} P_t^{k,l} \tag{8}$$

$$F_t^{cl} = \sum_{k=1}^{n_{cl}} \pi_t^{k,cl} P_t^{k,cl} \tag{9}$$

where $\pi_t^{k,pv}$, $\pi_t^{k,w}$ depict the offer prices by the kth PV panel and the wind turbine respectively; n_{pv}, n_{w}, n_{ES} indicate the number of PV panel, wind turbines and ES installed in the MG; $\pi_t^{k,ES-}$, $\pi_t^{k,ES+}$ represent the electricity cost of the kth storage unit during discharging and charging mode respectively; θ_{ES} is a binary number equal to 1 if the battery is charging and 0 otherwise; π_t^{G-}, π_t^{G+} depict the power price of the power imported from the grid and exported to the grid respectively; θ_{G} is a binary number equal to 1 if the MG is exporting to the grid and 0 otherwise; $\pi_t^{k,l}$, $\pi_t^{k,cl}$ represent the power cost of the of non-controllable loads and controllable load of the EWH respectively. n_{l}, n_{cl} indicate the number of the of non-controllable loads and controllable load; $P_t^{k,pv}$ is the power generated by the kth PV panel during time period t; $P_t^{k,w}$ is the power generated by the kth wind turbine during time period t; $P_t^{k,ES-}$, $P_t^{k,ES+}$ are the power generated and consumed of the kth storage unit during discharging and charging mode respectively; $P_t^{k,l}$ and $P_t^{k,cl}$ are the power generated and consumed by non-controllable loads and controllable load of the EWH respectively.

The objective function must be minimized subject to the following constraints:

- Power balance

$$\sum_{k=1}^{n_{pv}} P_t^{k,pv} + \sum_{k=1}^{n_w} P_t^{k,w} + \sum_{k=1}^{n_{ES}} \theta_{k,ES} \cdot P_t^{k,ES-} + \theta_G \cdot P_t^{G-}$$
$$= \sum_{k=1}^{n_{ES}} \left(1 - \theta_{k,ES}\right) \cdot P_t^{k,ES+} + (1 - \theta_G) \cdot P_t^{G+} + \sum_{k=1}^{n_l} P_t^{k,l} + \sum_{k=1}^{n_{cl}} \left(1 - \theta_{k,cl}\right) \cdot P_t^{k,cl+} \tag{10}$$
$$- \sum_{k=1}^{n_{cl}} \theta_{k,cl} \cdot P_t^{k,cl-}$$

- The renewable energy resources

$$0 \leq \sum_{k=1}^{n_{pv}} P_t^{k,pv} \leq P_t^{max,pv} \tag{11}$$

$$0 \leq \sum_{k=1}^{n_w} P_t^{k,w} \leq P_t^{max,w} \tag{12}$$

where $P_t^{max,pv}$, $P_t^{max,w}$ are the maximum available power from PV panels and wind generators during the period t.

- ES unit [28]:
  Maximum discharge limit:
$$\theta_{ES} \cdot P_t^{ES-} \leq P_{max}^{ES-}, \ P_t^{ES-} \geq 0 \tag{13}$$

  Maximum charge limit:
$$(1 - \theta_{ES}) \cdot P_t^{ES+} \leq P_{max}^{ES+}, \ P_t^{ES+} \geq 0 \tag{14}$$

  When the battery is in discharging mode $\theta_{ES} = 1$, the power discharged cannot go beyond the maximum discharging power of the battery P_{max}^{ES-}; similarly, the power charging the battery cannot go beyond the maximum power that could be consumed by the battery P_{max}^{ES+}.
  Maximum discharge limit considering the total stored energy:

$$(\theta_{ES} \cdot P_t^{ES-} \cdot \Delta t) \leq E_{t-1}^{ES} \tag{15}$$

- Maximum charge limit considering the total stored energy:

$$((1 - \theta_{ES}) \cdot P_t^{ES+} \cdot \Delta t) + E_{t-1}^{ES} \leq E_{max}^{ES} \tag{16}$$

- Energy balance in ESS:
$$E_t^{ES} = E_{t-1}^{ES} + (P_t^{ES+} - P_t^{ES-}) \cdot \Delta t \tag{17}$$

  The inequality (16) formulates the energy constraint for the battery in a way that the discharged should not exceed the available power. Similarly, the energy charging the battery should not exceed the upper battery limit. During the battery operation, the available energy E_t^{ES} should always be between the upper and lower limits E_{min}^{ES}, E_{max}^{ES} respectively as showed in Equation (18) [29].

$$E_{min}^{ES} \leq E_t^{ES} \leq E_{max}^{ES} \tag{18}$$

- State of charge of the battery:

$$SOC_t = \frac{E_t^{ES}}{E_{tot}^{ES}} \tag{19}$$

- The interconnection with the grid [30]:

$$(1 - \theta_G) \cdot P_t^{G+} \leq P_{max}^{G+}, \ P_t^{G+} \geq 0 \tag{20}$$

$$\theta_G: \; P_t^{G-} \leq P_{max}^{G-}, P_t^{G-} \geq 0 \tag{21}$$

P_{max}^{G-} is the maximum power that could be injected in the grid; P_{max}^{G+} is the maximum power that could be imported from the grid.

3. Proposed Algorithm for the Real-Time Optimization

Among classical techniques to solve the DED are the dynamic programming [31,32] or the Lagrangian relaxation [4,33,34]. However, most deterministic optimization techniques are not suited for real-time optimization since they need more time to reach the optimum. Heuristic techniques, on the other hand, may reach a sub-optimum solution, but their computational time is very competitive, which make them adequate for real-time application. Much progress was made in heuristic techniques that have proven their efficiency as competitive optimization methods in power system [3,5,7,35–37]. The artificial immune algorithm is inspired from the biological process of immune cells defense mechanism. The search process of this optimization approach is fast and robust, which, not like the particle swarm and the genetic algorithm, search for potential solutions in the overall shape space and don't fall into the trap of sub-optimality convergence [38]. This work proposes the T-Cell algorithm as a variant of immune system algorithm to solve the DED in real time MG operation.

The T-Cell algorithm is inspired by the mediated immune cells in the human body. These cells are called lymphocytes and develop in the Thymus as group of white blood cells [14]. They play a central role in cell-mediated immunity. They have receptors called T-Cell Receptors (TCR) that interact with proteins in their environment. Through this interaction, the T-Cells suffer two stages to get activated: proliferation then differentiation. In the first process, T-Cells proliferate by generating other clones of themselves; then, in the second process, each clone cell acquires new propriety; this is the so-called activation process [15]. This algorithm implements this process:

Each cell has nine characteristic values:

- P_t represents the decision variables of the dispatch problem;

$$P_t = \begin{bmatrix} P_t^{1,pv} & \cdots & P_t^{n_{pv},pv} \\ P_t^{1,w} & \cdots & P_t^{n_w,w} \\ P_t^{1,ES-} & \cdots & P_t^{n_{ES-},ES-} \\ P_t^{1,ES+} & \cdots & P_t^{n_{ES+},ES+} \\ P_t^{G-} & & \\ P_t^{G+} & & \\ P_t^{1,cl-} & \cdots & P_t^{n_{cl-},cl-} \end{bmatrix} \tag{22}$$

- The objective function value for P_t;
- The proliferation is the number of clones that the cell will generate;
- The differentiation is the number of decision variables that will be changed when the differentiation process takes place;
- The power generated by P_t;
- The equality constraint violation (ECV) for P_t ($|P - P_l|$);
- The inequality constraints violation (ICV) sum;
- The feasible propriety indicates if the cell is feasible or not. A cell is considered as feasible if: ECV = 0 and ICV = 0. This means that if a solution generates less than the demanded power (ECV < 0) or more than needed (ECV > 0), or the decision variable are in a prohibited zone (ICV < 0) or (ICV > 0) then it is considered as infeasible.

The required parameters for the T-Cell algorithm are: Number of objective function evaluations, population size and probability of the redistribution operators. The appropriate values of each parameter were chosen from the analysis result of this study [15].

Figure 1 describes the T-Cell algorithm execution steps.

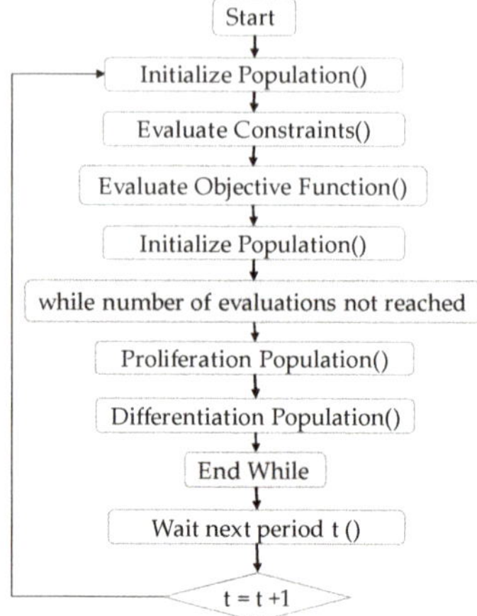

Figure 1. T-Cell algorithm process.

4. Proposed Multiagent EMS Platform

This section presents the MAS implemented for MG management and will present the developed agents and the proposed management strategy.

The agents are implemented in the Java Agent Development framework (JADE) platform. This platform respects the Foundation for Intelligent Physical Agents' (FIPA) standards that are considered as the reference for MAS standards used by computer science developers' community. JADE platform has the default Agent Management System (AMS) that controls the platform, creates and destroys other agents. The default Directory Facilitator (DF) is the agent responsible for registering all connected agents of the platform. Thus, if another agent wants to communicate with his peer, he has to ask the DF who will provide him with the name of the connected agent. In addition to these agents, the Agent Communication Channel also called the Message Transport System, is the agent in charge of transferring messages between agents of the platform [39]. JADE also provides graphical management service for platform management and control through the remote management agent (RMA). Any agent created in JADE is hosted in a container. This latter has all services for executing and hosting agents. Figure 2 presents the structure of the JADE platform, according to the FIPA agent management reference model.

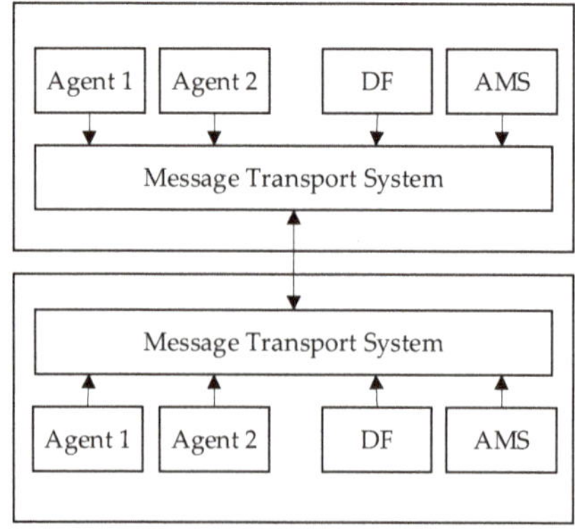

Figure 2. Jade platform architecture.

The MG considered in this work is composed of wind generators, PV panels, storage units, controllable load, and non-controllable load. The MG is operating in grid-connected mode. Each one of these components has a specific agent. There are three agent categories: The management agents, backup agent, and communication agents.

The MAS implemented is illustrated in Figure 3.

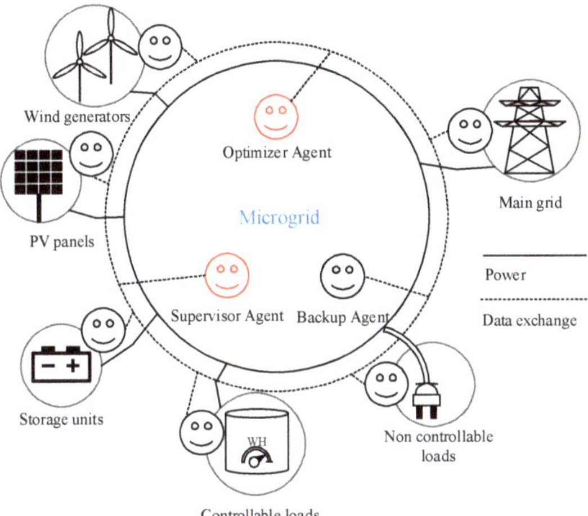

Figure 3. MAS architecture.

4.1. Management Agents

The first category comprises agents that ensure the proper operation of the MG. The MG optimizer (MGO) is the agent responsible for running the optimization and ensuring the seamless operation of the MG in real time. The T-Cell algorithm is implemented in this agent. This agent gets the forecast data of the wind, PV, and load power and computes the references of each generation unit and controllable load at each time step 15 min. However, during this period, an unpredictable load variation can occur, which can jeopardize the MG operation. So, instead of running the optimization every few seconds which is resource consuming, this work presents a solution to implement a specific agent that checks continually if the load power has gone beyond permissible limits during the time step.

The MG supervisor (MGS) is the load supervisor agent. When the MGS intercepts sudden high load variations, it contacts the MGO that launches a new optimization cycle and computes the new optimum operating points of the generators and controllable loads.

4.2. Backup Agent

The Main Container (MC) is the special node which the agent platform relies on. It coordinates between the other nodes and hosts the AMS and the DF. Since this node is very critical, if it terminates or fails to connect to other containers, some vital operational features can become unavailable, and this severely hampers the platform [40]. In order to strengthen the agent platform and make it more reliable, a fault tolerant structure is implemented. It can be observed from Figure 4 that there are two MC in the agent platform. Replication support for MC has been introduced. Using this support, it is also possible to duplicate as many as MC in the system. These MC arrange themselves in a logical ring so that whenever one of them fails, the others will notice and act accordingly [18].

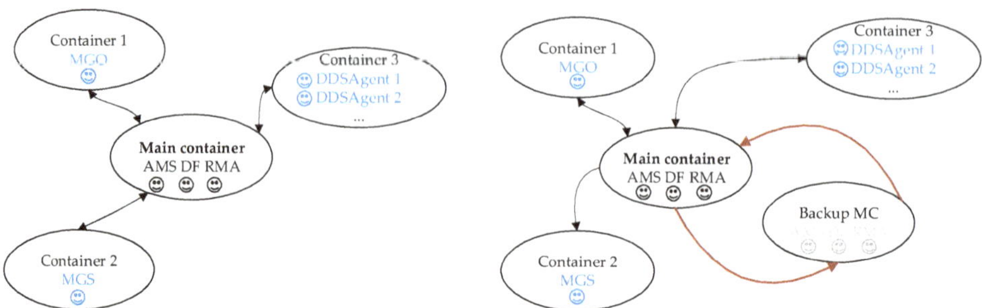

Figure 4. The fault tolerance mechanism in MAS platform.

4.3. Interoperability Agents

The main function of interoperability agents is to ensure the communication between the MAS platform and the physical hardware. The communication is based on the data-centric publish-subscribe (DCPS) model.

There are two main distributed communication models that could be implemented in MG context: Data-centric communications and object-centric communication model. In data-centric communications, data distribution between communicating entities is the fundamental concept. This model is based on data publishers and data subscribers. Publishers pass data of known types to subscribers through a common channel. In the opposite, in object oriented communications, instead of focusing on the data, the main concern is the interface between entities. An interface comprises of a list of predefined methods ready to be used. An object oriented system is based on server and client interface communication. The client interface invokes methods on a named server interface, which the server responds accordingly [41]. Even though the two approaches are complementary, middleware built on data-centric communication are more competitive in the RT applications.

RTI DDS Connect was implemented as middleware for RT interoperability between the agent context and the hardware of the testbed. As explained earlier, data-centric communication is based on data sharing between publishers and subscribers. The RTI DDS Connext allows to user to create publishers and subscribers independently of the programming language. Publishers and subscribers are connected by 'Topics.' A Topic corresponds to a single data type that allows identifying which data is being to be shared between which publishers and subscribers [42]. For instance, if publisher A is publishing value 'x' in Topic 1, subscribers who want to get the 'x' value should connect to the same Topic 1.

In this work, a publisher for each generation units and controllable load were implemented in the MAS context. The MGO computes the references and sends them to the corresponding publisher agents. Once the data is published, the subscribers implemented in the SCADA system of the testbed read the data and operate accordingly. Figure 5 illustrates how the agents implemented in the agent context publish data in RTI DDS topics, and how each generation unit reads the published data through subscribers.

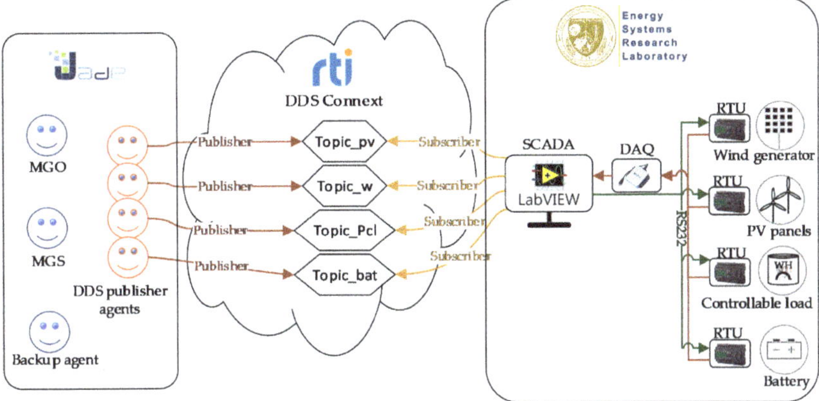

Figure 5. Interoperability Communication Setup.

5. Results

5.1. Description of the Cyber-Physical Setup

This section describes the implemented configuration of the testbed for the multi-agent RT-EMS experimentation in the Energy Systems Research Laboratory (ESRL), Florida International University.

The MG considered in this study is composed of a wind generator and PV system, each one is emulated by a synchronous generator, a storage unit emulated by a DC power supply, a controllable load, load, and connection to the grid.

Each generating station has five different types of motor drives, acting as five different prime movers to enable the implementation of various generation control strategies. These generators are equipped with frequency drives, Automatic Voltage Regulator (AVR) and synchronizer systems to control these different prime movers. The voltage regulators and automatic power control module controllers enable the system to operate as real-time generators. An automatic synchronizer connects each generator and controls the switching and measurement busses. The control and setup of each generation station are described in Figure 6.

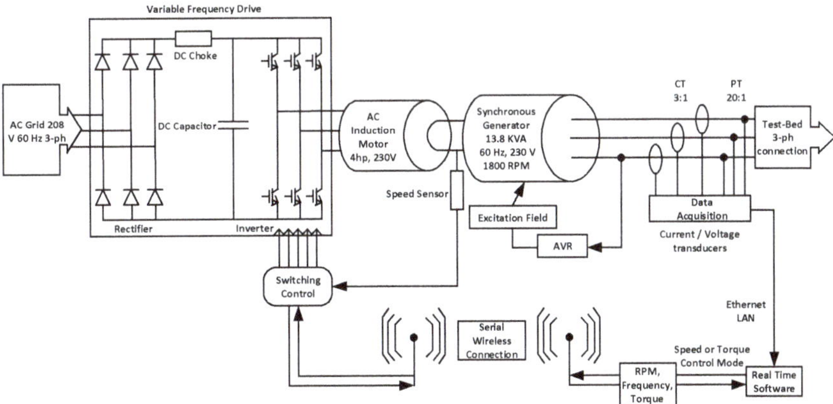

Figure 6. Generation station hardware and control.

The DC zone that emulates the DC storage unit is represented by a DC bus, where its voltage was set to 380 V and XR SERIES DC power supply offered by the MAGNA-POWER ELECTRONICS, which emulates the charging/discharging battery characteristics. It is interfaced with the DC bus through the DC-DC boost converter. The converter controllers are implemented in SIMULINK environment and executed in the DSpace 1104 real time interface. The system topology can be changed by using eight controlled switches. The switches have two states on and off. A control signal is sent to these switches to change their state which changes the equivalent circuit. The control is initiated from a virtual instrument (VI) program implemented in LabVIEW environment. The card PCI 6025E sends the control command to the circuit. This circuit is an inverting buffer module sn7406n of TEXAS INSTRUMENT. This module contains six inverters with open collector output. The RMS values for the voltage in the AC zone is set to 208 V. The AC bus is connected to the utility grid at PCC. Also, different load models were designed to represent the AC load pattern. Figure 7 shows the control of the power supply.

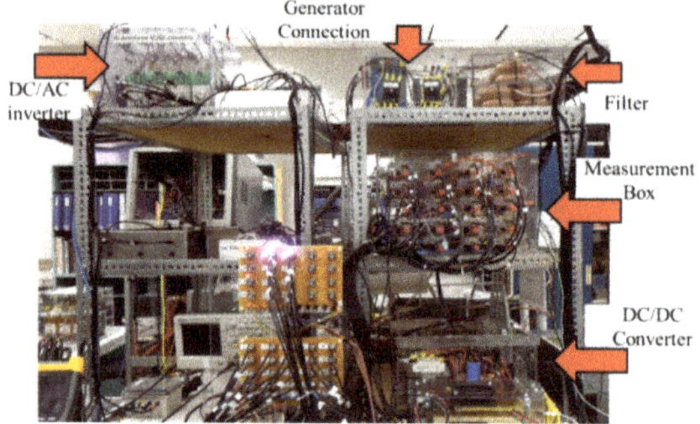

Figure 7. Control of DC power supply.

The testbed is controlled and monitored through the LabVIEW SCADA interface, as shown in Figure 8. The connected components for this experiment have switches with light green color.

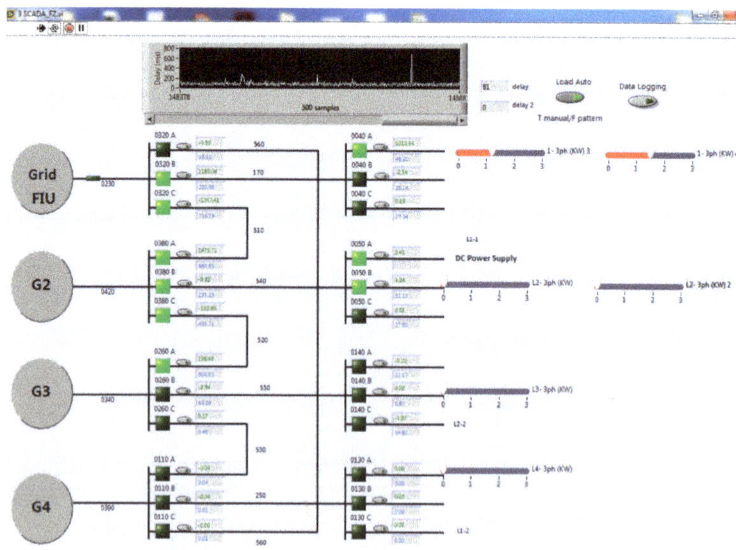

Figure 8. SCADA interface in ESRL.

5.2. Experiment Results

This section presents the results obtained from simulation and experimentation of the multi-agent RT-EMS. Three case studies are considered. The first case study presents the implemented load supervision mechanism of the MAS. The second case study describes the operation of the fault tolerance mechanism in the MAS. The third case study compares the results of the experimentation and the simulation and presents the competitivity of the implemented RT-EMS in terms of time delay.

5.2.1. Case Study 1

This case study presents the implemented RT load supervision mechanism. The MGO conducts the optimization every 15 min based on the forecasted PV, wind, and load data. However, due to some exceptional event, the load value can occasionally make a high variation from the forecast value, which affects the stability of the MG operation. For this reason, the MGS was conceived to supervise the high load variation in RT. The MGS is a cyclic agent that executes every 2 s. It computes the variation between the RT load and the forecast values. When the computed difference exceeds 30%, the MGS sends a signal to the MGO to start another optimization cycle based on the new sensed load value.

Figure 9 shows the communication mechanism implemented in Jade platform between the MGO, MGS, and other publisher agents.

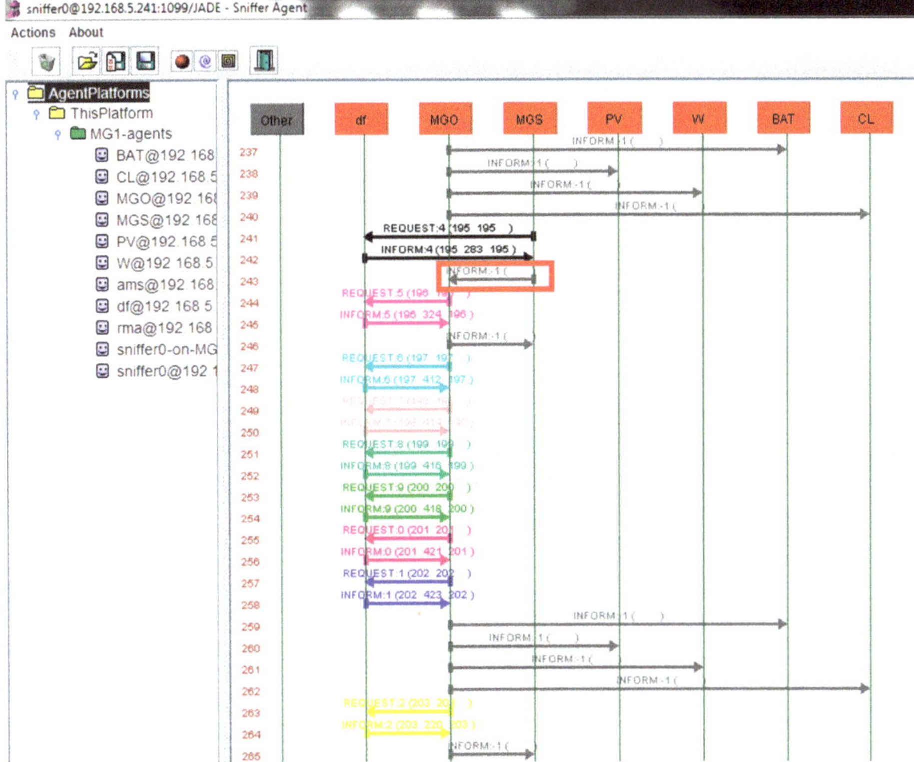

Figure 9. Communication between MGO, MGS and other agents.

At line 241, the MGS detects a high variation of the load. Then, the MGS sends a request to DF agent to send him references of connected MGO. Once credentials are delivered, the MGS send to the MGO the signal to start a new optimization cycle in line 243. An optimization communication cycle starts from line 244 to line 262. The MGO at the beginning of each cycle sends the parameters of the current cycle to the MGS. Then, the MGO launches the optimization based on the T-Cell algorithm. At the end of the optimization, the computed references are sent to the publisher agents.

5.2.2. Case Study 2

This case study describes the implemented operation of the fault tolerance mechanism in the MAS. As explained in Section 3, the Main Container (MC) is a special node in the JADE platform that coordinates between other containers and agents hosted in these containers. In order to implement a fault-tolerant mechanism, a backup agent of the MC was created. The two main-containers are arranged in a unidirectional ring. The MC is considered as the master MC while the backup agent is considered as the slave. Once the master MC fails, the slave MC detects the failure and takes the appropriate recovery actions. Three containers were created in this case study. By default, the AMS and DF agents are affected to the master MC, which is the 'Main-Container' in our case. The 'BackupContainer' is created, and 'Container-1' hosts the MG agents. 'Mg1' is the MGO while other agents are the interoperability agents of each MG components. Figure 10 illustrates the MAS platform before and after the fault test. A manual failure event was introduced while the RT-EMS was running. Through the Jade user interface, the MC was terminated. Automatically and instantaneously, the backup agent detects the event, changes its status to the new master MC and transfers the AMS and DF from the faulty MC to its container. As a consequence, the list of all alive agents and containers in the multi-agent platform remain updated and the RT-EMS continues its normal operation.

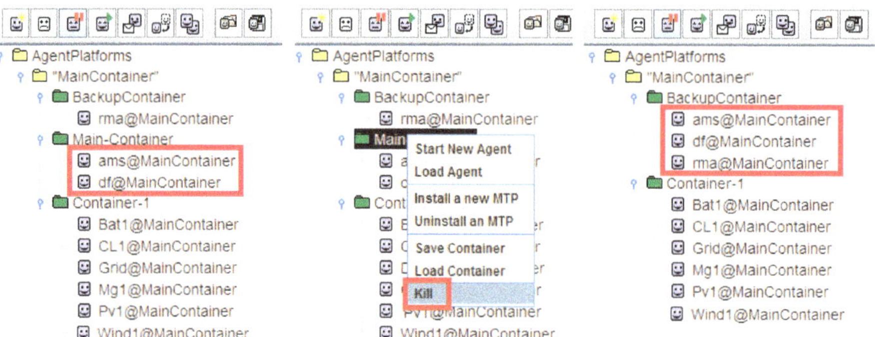

Figure 10. Fault tolerance mechanism in the JADE platform.

5.2.3. Case Study 3

This case study compares the results of the conducted experimentation and the simulation of the RT-EMS and presents the performance of this system compared to previous work.

The experimental validation of the implemented T-Cell based RT-EMS developed in a MAS environment was tested in the Energy Systems Research Laboratory at Florida International University. Three different scenarios were taken into consideration to validate the efficiency and performance of the proposed algorithm:

- Scenario 1: Normal operation of the MG. In this case, the system is running without sudden event. The MGO is computing and sending references of the generation units according to forecast data without any incident;
- Scenario 2: Generation drop. This case tests the ability of the system to operate after a significant shift in power generation. For this matter, the wind generator has been shut down during the period 6:00–7:00;

- Scenario 3: Sudden load variation. This scenario challenges MG operation when there is a sudden load increase. It is reflected in the period 15:00–16:00, where the load increases from 1500 to 2400 W.

Table 1 presents the parameters of the considered system. It presents the interval values of each system variable. The energy storage is considered operating between 0.4 and 0.9 to protect the battery from the deep discharging and overcharging.

Table 1. Microgrid parameters.

Variable	Meaning of the Variable	Limit Values
P_{pv}	The predicted power generation of the PV system	$0\text{ W} \leq P_{pv} \leq 700\text{ W}$
P_w	The predicted power generation of the wind turbine	$0\text{ W} \leq P_w \leq 1500\text{ W}$
P_{bat}	The predicted charging or discharging power of the battery	$-1500\text{ W} \leq P_{bat} \leq 1500\text{ W}$
SOC	State of charge of the battery	$0.2 \leq SOC \leq 0.9$
P_l	Active power load	$0\text{ W} \leq P_l \leq 2700\text{ W}$
P_{cl}	Controllable load	$P_{cl} = 0.2\, P_l$

Figures 11 and 12 illustrate the power profile of the wind and PV systems. As explained in Section 4.1, both wind and PV were emulated by two synchronous generators. It is noticeable from the two profiles that at the beginning of each period, there is a power fluctuation due to system transients. The power measured during the experiment is also slightly below simulated values because of the system losses. The wind generator was supposed shut down during the period 6:00–7:00 and the PV power profile was also assumed to be null before 7:00 and after 20:00; yet due to experimental constraints the generators were kept running at the minimum level. However, in Figures 13–15, the experimental profile of the energy storage, the load and the controllable load matches the simulated values. The wind and PV profiles data used in this experiment are available in [43] and [44], respectively.

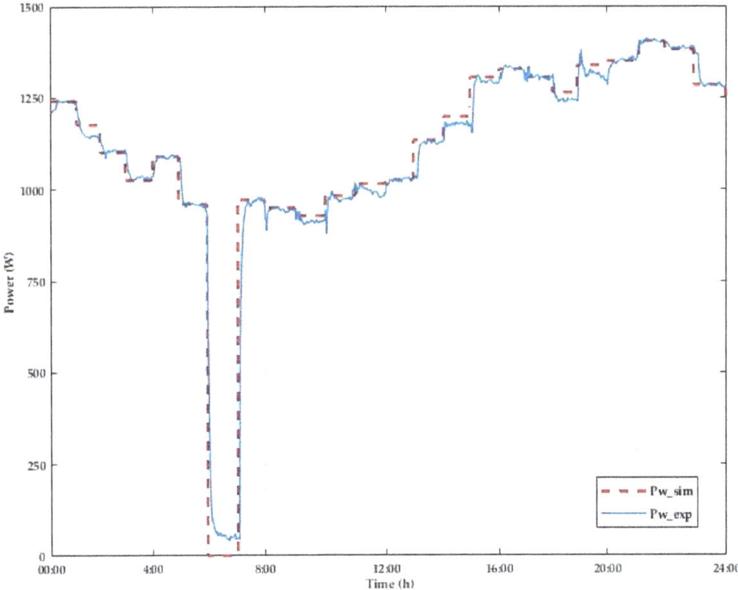

Figure 11. Comparison between simulated output and testbed measured values of wind power.

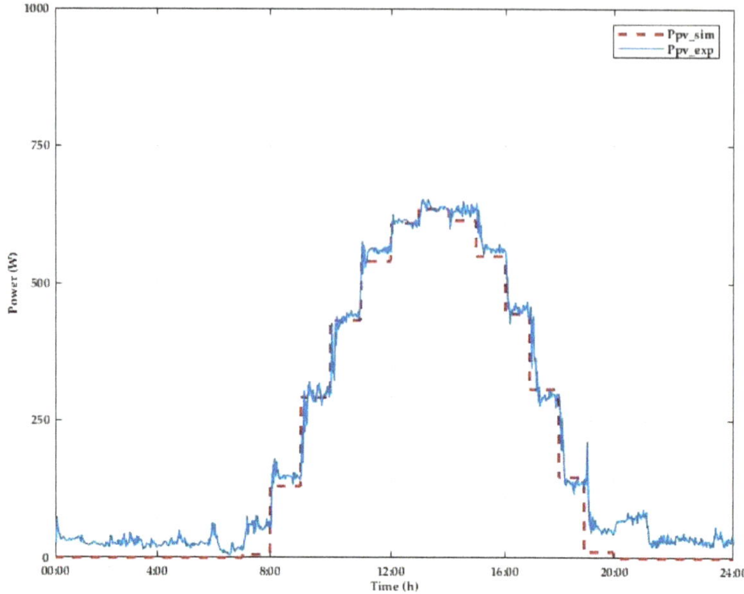

Figure 12. Comparison between simulated output and testbed measured values of PV power.

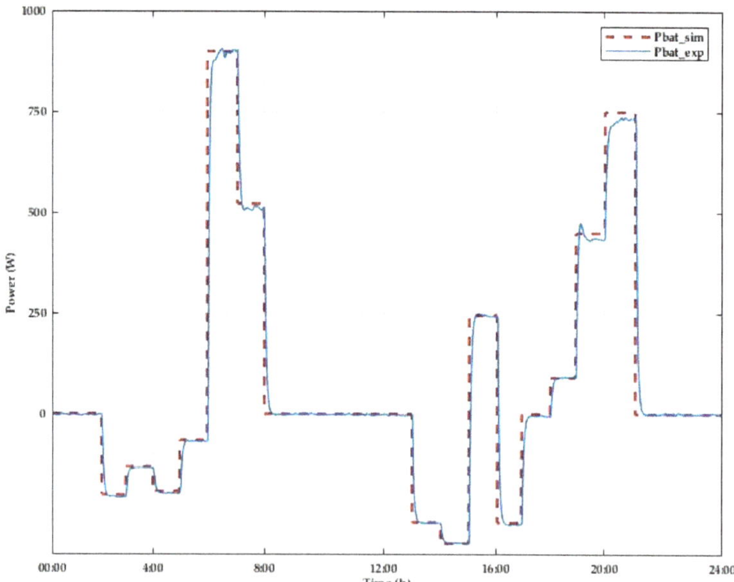

Figure 13. Comparison between simulated output and testbed measured values of energy storage power.

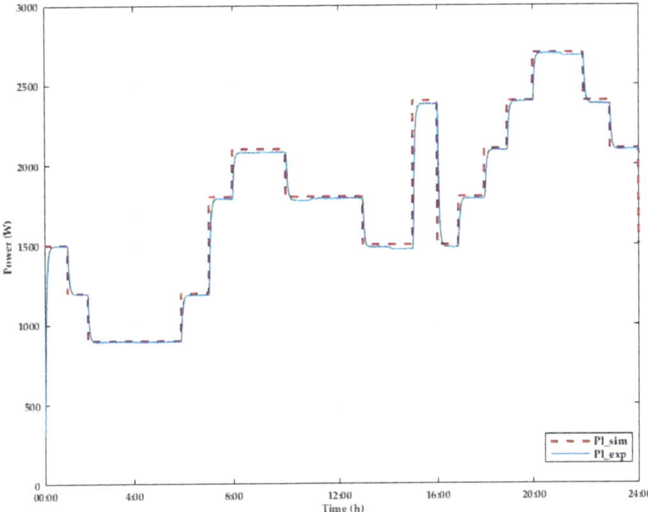

Figure 14. Comparison between simulated output and testbed measured values of load power.

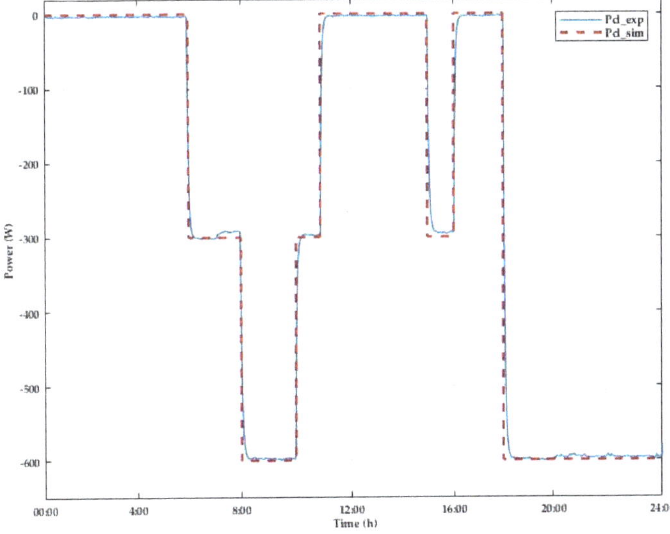

Figure 15. Comparison between simulated output and testbed measured values of controllable load.

Figure 16 describes the power exchange with the grid. Unlike the PV and the wind emulators that receive the reference from EMS, the power exchange with the grid is automatically imported or injected depending on the system consumption and generation. For this reason, all system errors are compensated by the grid power which creates, in some cases, a mismatch between the experiment and simulation values. It also should be mentioned that the simulation shows the RMS value without any transient, but the experimental result demonstrates the actual values along with transient of the power system which includes the compensation within the generators, reactions and power grid losses.

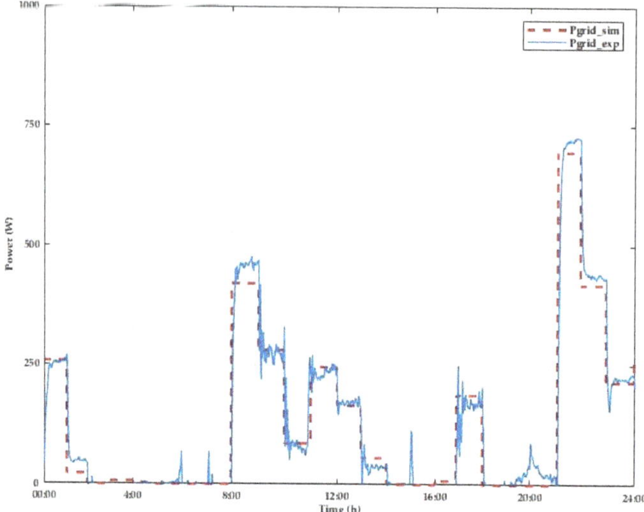

Figure 16. Comparison between simulated output and testbed measured values of grid power.

Figure 17 illustrates the MG dynamic during a day of operation going through the three scenarios. From 00:00 to 2:00, the MG is supplying the load from the wind and the grid. As soon as the load decreases from 2:00 to 6:00, the proposed algorithm allows charging of the energy storage with the excess power. During the shutdown of the wind generator from 6:00 to 7:00, reflected by a drop of the wind power (scenario 2), the needed power was supported by the energy storage and the controllable load. The following day's hours match more load increase, which is supplied by the grid, wind, controllable load, and the increasing PV power. From 13:00 to 15:00, the PV power is at the top generation, which allows charging the energy storage with excess power. Scenario 3 is tested from 15:00 to 16:00 in which the load suddenly raises from 1500 to 2400 W. The algorithm compensates the deficit power using the energy storage and controllable load. The MG continues its normal operation and uses the available power from the energy storage until it reaches its minimum value; then, the power needed is imported from the grid and the available wind power.

Through the 24 h of operation in different scenarios, the proposed T-cell based RT-EMS has proven his effectiveness and capability of maintaining the stability and balance of the MG.

The proposed T-Cell algorithm that has been implemented for the EMS optimization is coded in Java language using the Eclipse programming environment. The environment integrates the JADE tool kit and the DDS interoperability agents. Before implementing those agents, it is important to define the data model for the MG. In our case, the data model consists of four variables of type double that represents the reference computed by the MGO. These four variables are published through the DDS agents into four topics, as illustrated in Figure 18. In the other side, the LabVIEW SCADA system implements the subscribers for these topics. Once the new data is published, it is automatically intercepted by the subscribers and sent to the appropriate generator or controllable load.

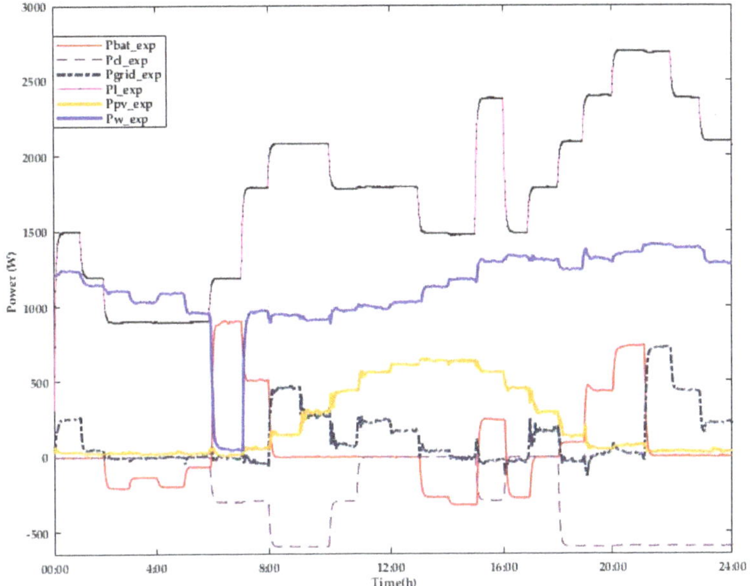

Figure 17. Experimental results of the MG powers during 24 h operation.

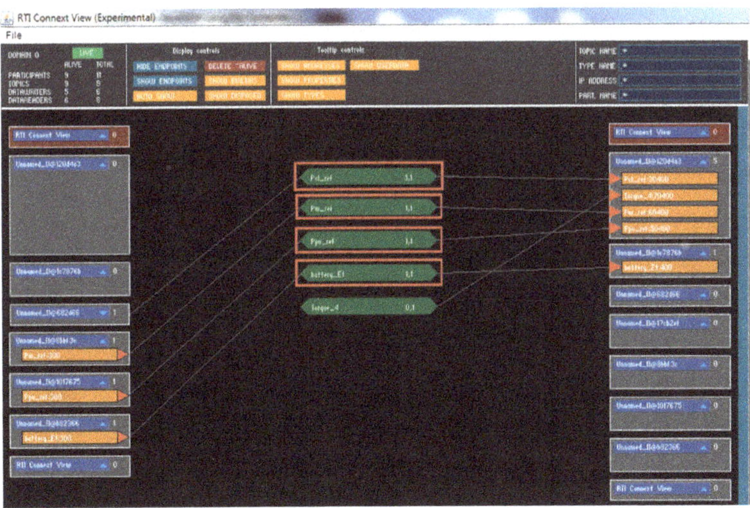

Figure 18. DDS Connext view.

The RT-EMS has been tested using a computer with the following specifications: 4GB of RAM and 2 GHz of CPU. In this configuration, the average execution time of the T-Cell algorithm is 231 ms. However, when executed in a computer of 6G of RAM, the average execution time is 60 ms. In both cases, the T-Cell algorithm presented better performance than mixed integer nonlinear programming (MINLP) and multi-period artificial bee colony algorithm (MABC) that have 8.23 s and 1.14 s respectively [10].

The execution time is very crucial for RT-EMS in MG context and not as important as in day ahead management where the optimization is run a day before. The notion of real time that has been reported in different papers is between 3 to 5 min. The reason behind choosing this time is because it is less than the time needed for the adjustment of protective relay such as over/under voltage relay. Besides, PV systems and wind generators usually need more than 5 s to be affected by atmospheric variations. Although the implemented RT-EMS could be executed every 5 min and still presents better results than previous works, it is executed every 15 min. The main reason is that the system has the MGS, which is continually checking for high load variation and in this case, launches a new optimization cycle. This way, system resources are used optimally. The MG under study has an RS232 serial connection network for the control of the testbed components and the SCADA system, as illustrated in Figure 5. The latency time for sending data from the LabVIEW SCADA system to the generators and controllable loads in the testbed is 1 s [45].

Furthermore, for a measurement message size of 32 bytes and message rate of 1000 Msg/s, the average latency for DDS middleware is 243 µs with 90% below 269 µs [46]. The T-Cell optimization algorithm executed in a computer of 4GB of RAM and 2 GHz of CPU has an average latency of 231 ms. Therefore, the total latency of the RT-EMS is 1.3 s on average, which is way less than the 3 min presented in [10].

Another positive point could be stated. As the MG size increases and additional components are connected, the system variables will increase. This would be reflected in the execution time that would increase too. However, the proposed immune algorithm was tested along with fast, chaotic particle swarm optimization (PSO) and independent component analysis based PSO in the case of the small and medium power network, and the T-Cell algorithm reached a better optimum in minimum time [15].

5.3. Discussion

The proposed RT-EMS has several features that make it easy to implement in MGs as well as in smart buildings. The MAS makes it easier to extend the system and add other components. New agents can be designed for new components and integrated in the agent platform and then added in the optimization system. In addition, the DDS makes it very easy to integrate any type of hardware to the existing MG. For instance, a new PV system in a smart building, can interfaced to the SCADA using a new publisher and a new topic for this new PV system. Thus, the RT-EMS is modular and easy to extend for larger MGs.

Nevertheless, the system can be improved by considering the distribution losses and constraints in the optimization. Since the present work focuses on small scale MGs e.g., smart home, where all generations and consumption units are in the same location, the losses are not significant; for this reason, this assumption was made. However, one of the future improvements of the present work is to take it into consideration. The authors considered the water heater as a controllable load as it can be easily turn on and off without affecting costumer comfort. Besides, the relatively small size of the controllable power used in this work make it comparable to a water heater. However, in a large Mg, the controllable loads could be considered as the centralized air conditioner, the washing machine or dishwasher which present a bigger power to control. A second optimization layer between MGOs that negotiate the best option for buying or selling the power depending on the market prices could be added. Besides, the grid connected mode was considered in this study. Future works can consider the islanded mode of operation that updates the RT-EMS parameters accordingly.

6. Conclusions

Smart cities are no longer a future vision but a reality. Technologies are not only shaping our modern life but also helping us to preserve the environment and adopt a more sustainable life. Several cities around the world launch programs aiming to preserve the environment. While some cities choose to promote the use of applications to ease the city traffic and therefore reduce gas emissions, others present incentives to its citizens to produce renewable energy and reduce their consumption during peak hours; In this context, developing a comprehensive EMS that would enhance cities sustainability agenda is a hot topic. Our proposed solution that incorporate fast real time optimization would be a good asset for sustainable cities. A multiagent RT-EMS based on T-Cell algorithm was implemented and successfully experimented in the smart grid testbed at Florida International University for optimally managing a MG. Three scenarios were tested to show the effectiveness and optimal operation of the MG using the proposed technique. Comparison of simulations and experimental results show the ability of the proposed multiagent T-Cell based RT-EMS in maintaining the stability and smooth operation of the MG. In addition, the modularity and fault tolerance features were easier to implement through the MAS JADE platform. The DDS middleware also allows the interoperability between different components of the system. The results obtained by the multiagent RT-EMS highlight the effectiveness of the proposed algorithm and demonstrate a faster convergence time compared to previous works. The optimum values were obtained faster in terms of computation time as compared to existing techniques. The latency from the proposed system was 43% faster than other heuristic or deterministic methods in the literature. This significant improvement makes this proposed system more competitive for RT applications. Therefore, our proposed EMS could be envisioned as a real solution for future RT operating MG. Furthermore, this EMS could also be implemented in smart homes or smart buildings that integrate RE and need RT EMS. Through the use of RT-EMS, the MG operator not only guaranties reduction of energy consumption cost, but also, assure the maximum utilization of available RE. As a consequence, the system is more resilient and participate in reducing the negative environmental impact. The RT-EMS also apply DR when needed, which have a positive impact on the distribution system during peak hours.

Author Contributions: F.Z.H. conceived, implemented the software and wrote the original draft. A.F.E. and M.M.E. contributed to the experimental, review and editing work. N.K. is the responsible for funding acquisition and contributed to work investigation and supervision. N.H. contributed with materials tools and supervision. O.A.M. is the supervisor who leads the project and edits the manuscript and he is the corresponding author.

Funding: This research was funded by IRESEN in the framework of the Inno-PV research project "SECRETS—Sustainable Energy Clusters Realized Through Smart Grids" and the Fulbright Join Supervision program.

Acknowledgments: This work was supported by Energy Systems Research Laboratory, Department of Electrical and Computer Engineering, Florida International University, Miami, Florida USA.

Conflicts of Interest: The authors declare no conflict of interest.

References

1. Margoum, E.H.; Krami, N.; Harmouch, F.Z.; Al montaser, H.; Seca, L.; Moreira, C. Design and Control of Single Phase Photovoltaic Systems for AC MicroGrid. In Proceedings of the 2016 International Renewable and Sustainable Energy Conference (IRSEC), Marrakech, Morocco, 14–17 November 2016; pp. 1188–1193. [CrossRef]

2. Margoum, E.H.; Krami, N.; Seca, L.; Moreira, C.; Mharzi, H. Design and Control of Parallel Three Phase Voltage Source Inverters in Low Voltage AC Microgrid. *Adv. Electr. Electron. Eng.* **2017**, *15*, 120–129. [CrossRef]

3. Jebaraj, L.; Venkatesan, C.; Soubache, I.; Rajan, C.C.A. Application of Differential Evolution Algorithm in Static and Dynamic Economic or Emission Dispatch Problem: A Review. *Renew. Sustain. Energy Rev.* **2017**, *77*, 1206–1220. [CrossRef]

4. Dieu, V.N.; Ongsakul, W.; Polprasert, J. The Augmented Lagrange Hopfield Network for Economic Dispatch with Multiple Fuel Options. *Math. Comput. Model.* **2013**, *57*, 30–39. [CrossRef]

5. Secui, D.C. Large-Scale Multi-Area Economic/Emission Dispatch Based on a New Symbiotic Organisms Search Algorithm. *Energy Convers. Manag.* **2017**, *154*, 203–223. [CrossRef]

6. Tade, S.V.; Ghate, V.N.; Mulla, S.Q.; Kalgunde, M.N. Application of Dynamic Programming Algorithm for Thermal Unit Commitment with Wind Power. In Proceedings of the 2018 IEEE Global Conference on Wireless Computing and Networking, GCWCN, Lonavala, India, 23–24 November 2018; pp. 182–186. [CrossRef]

7. Mohsen, N.; Braun, M.; Tenbohlen, S. Optimization of Unit Commitment and Economic Dispatch in Microgrids Based on Genetic Algorithm and Mixed Integer Linear Programming. *Appl. Energy* **2018**, *210*, 944–963.

8. Moghaddam, M.M.; Marzband, M.; Azarinejadian, F. Optimal Energy Management for a Home Microgrid Based on Multi-Period Artificial Bee Colony. In Proceedings of the 25th Iranian Conference on Electrical Engineering (ICEE2017), Tehran, Iran, 2–4 May 2017; pp. 1446–1451.

9. Shi, W.; Li, N.; Chu, C.-C.; Gadh, R. Real-Time Energy Management in Microgrids. *IEEE Trans. Smart Grid* **2017**, *8*, 228–238. [CrossRef]

10. Marzband, M.; Ghazimirsaeid, S.S.; Uppal, H.; Fernando, T. A Real-Time Evaluation of Energy Management Systems for Smart Hybrid Home Microgrids. *Electr. Power Syst. Res.* **2017**, *143*, 624–633. [CrossRef]

11. Netto, R.S.; Ramalho, G.R.; Bonatto, B.D.; Carpinteiro, O.A.S.; Zambroni De Souza, A.C.; Oliveira, D.Q.; Braga, R.A.S. Real-Time Framework for Energy Management System of a Smart Microgrid Using Multiagent Systems. *Energies* **2018**, *11*, 656. [CrossRef]

12. Aragon, V.S.; Esquivel, S.C.; Coello Coello, C.A. Optimizing Constrained Problems through a T-Cell Artificial Immune System. *J. Comput. Sci. Technol.* **2008**, *8*, 158–165.

13. Alberts, B.; Johnson, A.; Lewis, J.; Raff, M.; Roberts, K.; Walter, P. Helper T Cells and Lymphocyte Activation. In *Molecular Biology of the Cell*; Garland Science: New York, NY, USA, 2002.

14. Aragon, V.S.; Esquivel, S.C.; Coello Coello, C.A. A T-Cell Algorithm for Solving Dynamic Optimization Problems. *Inf. Sci.* **2011**, *181*, 3614–3637. [CrossRef]

15. Aragón, V.S.; Esquivel, S.C.; Coello Coello, C.A. An Immune Algorithm with Power Redistribution for Solving Economic Dispatch Problems. *Inf. Sci.* **2015**, *295*, 609–632. [CrossRef]

16. Harmouch, F.Z.; Krami, N.; Benhaddou, D.; Hmina, N.; Zayer, E.; Margoum, E.H. Survey of Multiagents Systems Application in Microgrids. In Proceedings of the 2nd International Conference on Electrical and Information Technologies ICEIT'2016, Tangier, Morocco, 4–7 May 2016; pp. 270–275.

17. Mcarthur, S.D.J.; Davidson, E.M.; Catterson, V.M.; Dimeas, A.L.; Hatziargyriou, N.D.; Ponci, F.; Funabashi, T. Multi-Agent Systems for Power Engineering Applications—Part I: Concepts, Approaches, and Technical Challenges. *IEEE Trans. Power Syst.* **2007**, *22*, 1743–1752. [CrossRef]

18. Harmouch, F.Z.; Krami, N.; Hmina, N. A Multiagent Based Decentralized Energy Management System for Power Exchange Minimization in Microgrid Cluster. *Sustain. Cities Soc.* **2018**, *40*, 416–427. [CrossRef]

19. Pinto, T.; Ghazvini, M.A.F.; Soares, J.; Faia, R.; Corchado, J.M.; Castro, R.; Vale, Z. Decision Support for Negotiations among Microgrids Using a Multiagent Architecture. *Energies* **2018**, *11*, 2526. [CrossRef]

20. Cintuglu, M.H.; Martin, H.; Mohammed, O.A. Real-Time Implementation of Multiagent-Based Game Theory Reverse Auction Model for Microgrid Market Operation. *IEEE Trans. Smart Grid* **2015**, *6*, 1064–1072. [CrossRef]

21. Mahmoudian Esfahani, M.; Hariri, A.; Mohammed, O.A. A Multiagent-Based Game-Theoretic and Optimization Approach for Market Operation of Multi-Microgrid Systems. *IEEE Trans. Ind. Inform.* **2018**, *3203*, 1–12. [CrossRef]

22. Logenthiran, T.; Srinivasan, D.; Khambadkone, A.M. Multi-Agent System for Energy Resource Scheduling of Integrated Microgrids in a Distributed System. *Electr. Power Syst. Res.* **2011**, *81*, 138–148. [CrossRef]

23. Pinto, T.; Faia, R.; Navarro-Caceres, M.; Santos, G.; Corchado, J.M.; Vale, Z. Multi-Agent-Based CBR Recommender System for Intelligent Energy Management in Buildings. *IEEE Syst. J.* **2019**, *13*, 1084–1095. [CrossRef]

24. Anvari-Moghaddam, A.; Rahimi-Kian, A.; Mirian, M.S.; Guerrero, J.M. A Multi-Agent Based Energy Management Solution for Integrated Buildings and Microgrid System. *Appl. Energy* **2017**, *203*, 41–56. [CrossRef]

25. Bui, V.; Hussain, A.; Kim, H. Optimal Operation of Microgrids Considering Auto-Configuration Function Using Multiagent System. *Energies* **2017**, *10*, 1484. [CrossRef]

26. Choi, I.S.; Hussain, A.; Bui, V.H.; Kim, H.M. A Multi-Agent System-Based Approach for Optimal Operation of Building Microgrids with Rooftop Greenhouse. *Energies* **2018**, *11*, 1876. [CrossRef]

27. Gómez-Expósito, A.; Conejo, J.A.; Cañizares, C. *Electric Energy Systems Analysis and Operation*; CRC Press: Boca Raton, FL, USA, 2009.

28. Marzband, M.; Ghadimi, M.; Sumper, A.; Domínguez-garcía, J.L. Experimental Validation of a Real-Time Energy Management System Using Multi-Period Gravitational Search Algorithm for Microgrids in Islanded Mode. *Appl. Energy* **2014**, *128*, 164–174. [CrossRef]

29. Xiaoping, L.; Ming, D.; Jianghong, H.; Pingping, H.; Yali, P. Dynamic Economic Dispatch for Microgrids Including Battery Energy Storage. *IEEE Int. Symp. Power Electron. Distrib. Gener. Syst.* **2010**, *2*, 914–917. [CrossRef]

30. Marzband, M.; Parhizi, N.; Savaghebi, M.; Guerrero, J.M. Distributed Smart Decision-Making for a Multimicrogrid System Based on a Hierarchical Interactive Architecture. *IEEE Trans. Energy Convers.* **2016**, *31*, 637–648. [CrossRef]

31. Ross, D.; Kim, S. Dynamic Economic Dispatch of Generation. *IEEE Trans. Power Appar. Syst.* **1980**, *99*, 2060–2068. [CrossRef]

32. Liang, Z.X.; Glover, J.D. A Zoom Feature for a Dynamic Programming Solution to Economic Dispatch Including Transmission Losses. *IEEE Trans. Power Syst.* **1992**, *7*, 544–550. [CrossRef]

33. Hemamalini, S.; Simon, S.P. Dynamic Economic Dispatch Using Maclaurin Series Based Lagrangian Method. *Energy Convers. Manag.* **2010**, *51*, 2212–2219. [CrossRef]

34. Li, Z.; Wu, W.; Zhang, B.; Sun, H.; Guo, Q. Dynamic Economic Dispatch Using Lagrangian Relaxation With Multiplier Updates Based on a Quasi-Newton Method. *IEEE Trans. Power Syst.* **2013**, *28*, 4516–4527. [CrossRef]

35. Shilaja, C.; Ravi, K. Optimization of Emission/Economic Dispatch Using Euclidean Affine Flower Pollination Algorithm (EFPA) and Binary FPA (BFPA) in Solar Photo Voltaic Generation. *Renew. Energy* **2017**, *107*, 550–566. [CrossRef]

36. Chen, C.; Duan, S.; Cai, T.; Liu, B.; Hu, G. Smart Energy Management System for Optimal Microgrid Economic Operation. *IET Renew. Power Gener.* **2011**, *5*, 258. [CrossRef]

37. Faia, R.; Pinto, T.; Vale, Z.; Corchado, J.M.; Soares, J.; Lezama, F. Genetic Algorithms for Portfolio Optimization with Weighted Sum Approach. In Proceedings of the 2018 IEEE Symposium Series on Computational Intelligence, Bangalore, India, 18–21 Novermber 2018; pp. 1823–1829. [CrossRef]

38. Oliveira, D.Q.; Zambroni de Souza, A.C.; Almeida, A.B.; Lima, I. An Artificial Immune Approach for Service Restoration in Smart Distribution Systems. In Proceedings of the 2015 IEEE PES Innovative Smart Grid Technologies Latin America (ISGT LATAM), Montevideo, Uruguay, 5–7 October 2015; pp. 1–6.

39. Bellifemine, F.; Caire, G.; Greenwood, D. *Developing Multi-Agent with JADE Systems*; Wooldridge, M., Ed.; John Wiley & Sons: London, UK, 2007. [CrossRef]

40. Bellifemine, F.; Caire, G.; Trucco, T.; Rimassa, G.; Mungenast, R. *JADE Administrator's Guide*; TILab: Turin, Italy, 2010.

41. RTI DDS. What Is DCPS? Available online: https://community.rti.com/static/documentation/connext-dds/5.2.0/doc/manuals/connext_dds/html_files/RTI_ConnextDDS_CoreLibraries_UsersManual/Content/UsersManual/What_is_DCPS_.htm (accessed on 15 June 2019).

42. RTI DDS. Data Topics—What Is the Data Called? Available online: https://community.rti.com/static/documentation/connext-dds/5.2.0/doc/manuals/connext_dds/html_files/RTI_ConnextDDS_CoreLibraries_UsersManual/index.htm#UsersManual/Data_Topics___What_is_the_Data_Called_.htm%3FTocPath%3DPart%25201%253A%2520Introduction%7CDa (accessed on 15 June 2019).

43. The Pennsylvania State Climatologist. 2018. Available online: http://climate.psu.edu/ (accessed on 22 June 2019).

44. National Solar Radiation Data Base. 1991. Available online: https://rredc.nrel.gov/solar/old_data/nsrdb/1991-2010/hourly/siteonthefly.cgi?id=722020 (accessed on 22 June 2019).

45. Salehi, V.; Mohamed, A.; Member, S.; Mazloomzadeh, A.; Mohammed, O.A. Laboratory-Based Smart Power System, Part I: Design and System Development. *IEEE Trans. Smart Grid* **2012**, *3*, 1394–1404. [CrossRef]

46. Youssef, T.A.; Elsayed, A.T.; Mohammed, O.A. Data Distribution Service-Based Interoperability Framework for Smart Grid Testbed Infrastructure. *Energies* **2016**, *9*, 150. [CrossRef]

Article

Reactive Power Management Considering Stochastic Optimization under the Portuguese Reactive Power Policy Applied to DER in Distribution Networks

Tiago Abreu [1], Tiago Soares [1,*], Leonel Carvalho [1], Hugo Morais [2], Tiago Simão [3] and Miguel Louro [3]

1 INESC TEC - INESC Technology and Science, Centre for Power and Energy Systems,
   4200-465 Porto, Portugal; tiago.j.abreu@inesctec.pt (T.A.); leonel.m.carvalho@inesctec.pt (L.C.)
2 INESC-ID, Department of Electrical and Computer Engineering, Instituto Superior Técnico-IST,
   Universidade de Lisboa, 1049-001 Lisbon, Portugal; hugo.morais@tecnico.ulisboa.pt
3 EDP Distribution; 1050-121 Lisbon, Portugal; TiagoFilipe.Simao@edp.pt (T.S.); Miguel.Louro@edp.pt (M.L.)
* Correspondence: tiago.a.soares@inesctec.pt

Received: 1 October 2019; Accepted: 21 October 2019; Published: 23 October 2019

Abstract: Challenges in the coordination between the transmission system operator (TSO) and the distribution system operator (DSO) have risen continuously with the integration of distributed energy resources (DER). These technologies have the possibility to provide reactive power support for system operators. Considering the Portuguese reactive power policy as an example of the regulatory framework, this paper proposes a methodology for proactive reactive power management of the DSO using the renewable energy sources (RES) considering forecast uncertainty available in the distribution system. The proposed method applies a stochastic sequential alternative current (AC)-optimal power flow (SOPF) that returns trustworthy solutions for the DSO and optimizes the use of reactive power between the DSO and DER. The method is validated using a 37-bus distribution network considering real data. Results proved that the method improves the reactive power management by taking advantage of the full capabilities of the DER and by reducing the injection of reactive power by the TSO in the distribution network and, therefore, reducing losses.

Keywords: decision-aid; distributed energy resources; distribution system operator; reactive power management; uncertainty

1. Introduction

The implementation of renewable energy sources (RES) and the deployment of distributed energy resources (DER) have created a trend of evolution in the distribution network that requires the adaptation of the conventional practices to handle the behavior that is related to the RES [1]. These new procedures are compelled to have a more proactive role by the distribution system operator (DSO), controlling and/or contracting DER to deal with voltage and line/transformers congestions problems [2]. Incorporating forecast in the system operation of the DSO, as well as creating contract services with the DER, where the power flexibility is enabled to change the expected operating point, can contribute to resolve the network operational problems [3]. The power flexibility may be divided for active or reactive power, which can assist in the network problems at a certain cost. This will ensure that the DSO can maintain the ability of granting network access to consumers and producers, with power quality, safety, and stability.

The transmission system operator (TSO) and DSO coordination may be a path to explore as DSO finds here the opportunity to coordinate a reactive power service with the TSO. This coordination intends to avoid voltage and/or congestion problems in the transmission system assuring that the distribution continues working without problems. Depending on the policies and agreements, this service can be remunerated given an extra advantage to the DERs. In fact, preventive reactive power management models are emerging as a potential solution for improving the coordination between the DSO and the TSO, while ensuring proper levels of voltage control in the system, as shown in [4].

There are many literature examples regarding approaches on reactive power management with fixed active power injection as in [5,6]. These, however, do not consider networks with strong RES penetration. A stochastic approach for ensuring voltage stability is proposed in [7]. The method considers a two-stage stochastic model with multi-objectives, such as minimization of power losses, operation, and management costs and wind power costs. Similarly, [8] proposes a stochastic model for corrective voltage control under severe contingencies, considering the uncertainty of wind power producer and consumer demand. A coordinated active and reactive optimization of an active distribution network considering energy storage systems and relaxed optimal power flow is proposed in [9]. It proposes a multi-objective function for minimizing power losses, operation costs and voltage deviation, however, the reactive power provision to assist in the TSO/DSO coordination is disregarded.

Most of these works disregard the full behavior of the distribution grid, introducing approximations and linearized versions of the full alternative current optimal power flow (AC-OPF). This can lead to sub-optimal solutions that may be infeasible. Thus, [10] proposes a voltage sensitivity analysis for adjusting the reactive power setpoint of DER in order to improve voltage stability and provide reactive power to upper levels of the network. Complementarily, [11] models an adaptive control of the reactive power setpoints of wind farms to assist the TSO/DSO coordination, minimizing the losses while ensuring proper level of reactive power provision. Still, none of these works can schedule in advance adequate reactive power setpoints for DER, considering the uncertain and variable behavior of RES.

In this scope, the main objective of this paper is to propose a stochastic reactive power management model to assist the DSO in the reactive power management ahead of the operating hour. The main contributions of this paper are threefold:

- To design a two-stage stochastic reactive power management model considering a full AC-OPF. It has the purpose of aiding the decision-making of the DSO under the uncertain and variable behavior of RES connected in the distribution network;
- To propose a reactive power service provided by the DSO to the TSO in advance of the operating hour. This service can be used by the TSO in the transmission system management, defining a reactive power operation in the TSO/DSO boundary substations. This can help the TSO in different services like the voltage control and congestion management in the transmission system;
- Take into account the Portuguese reactive power policy on distribution grids, assessing the behavior and applicability of the proposed model.

This paper is structured as follows: Section 2 describes the Portuguese reactive power policies and introduces the sequential AC-optimal power flow (SOPF) tool model; Section 3 presents the mathematical formulation of the stochastic approach for reactive power management; Section 4 validates the proposed model based on a 37-bus distribution network with real data; Section 5 presents the most important conclusions.

2. Reactive Power Policies

2.1. Portuguese Reactive Power Policy

The Portuguese reactive power policy for the distribution network is based on the total inductive and capacitive reactive power that a generating unit produces in an hour [12]. The reactive power is dependent on the active power injected by the generating unit in the form of $\tan \phi$. The reactive power must have a deviation of less than +/−5% from the defined $\tan \phi$.

Each day is divided into four periods: peak, full hours, valley and super valley. Yet, reactive power has only two classifications: peak and off-peak. There are two different schemes referring to the generating units. The ordinary scheme encompasses conventional units, whereas the special scheme, RES, industrial and urban waste, cogeneration and micro-producers. The ordinary scheme is limited to $\tan \phi = 0.4$ for the peak period and $\tan \phi = 0$ for off-peak hours.

For the special scheme, Table 1 illustrates the relationship between active and reactive power.

Table 1. Reactive power policy for the special scheme [12].

Voltage Level	*tan* ϕ	
	Peak Period	**Off-Peak Period**
High Voltage	0	0
Medium Voltage (P > 6MW)	0	0
Medium Voltage (P ≤ 6MW)	0.3	0
Low Voltage	0	0

Similarly, there is a contractual agreement between the TSO and the DSO whereby the upstream connection will have a profile of $\tan \phi$ as regulated for the special scheme. This means that between seven–22 h the $\tan \phi$ should be within −0.3 and 0.3 and in the remaining hours $\tan \phi = 0$, with a deviation of less than +/−5% of the $\tan \phi$. Failure to do so results in a penalty applied to the entity responsible for the failure. This penalty is modelled in steps of $\tan \phi$ infringement for peak hours. Equation (1) illustrates the current 3 steps violation of the limits penalties [13,14]:

$$
\begin{aligned}
0.3 \leq \tan \phi < 0.4, \\
0.4 \leq \tan \phi < 0.5, \\
0.5 \leq \tan \phi
\end{aligned}
\tag{1}
$$

with penalty factor applied to the reference price of reactive power of 0.33, 1 and 3 for each respective step.

2.2. Proactive Reactive Power Management

Though DERs are applying current reactive power policies, the distribution system does not take full advantage of DER's technical capabilities (especially RES) or consider their variable and intermittent behavior. More precisely, the uncertain DERs production is often overlooked, and their reactive power contribution is as well. This may lead to difficulties in reactive power management, especially in distribution networks with a high level of DER integration. In this scope, the SOPF tool proposed in this paper utilizes a two-stage stochastic reactive power management model (presented in Figure 1) to account for the uncertainty and variability related to the DER. The model has been developed for single period simulation using information from a representative set of scenarios for the DER. Stochastic optimization is used as a means to handle the uncertainty of DER. Such a probabilistic approach is integrated into the reactive power management problem through scenarios of active power generation of the DER. As DER's active power generation is uncertain, reactive power can be as well. Still, reactive power generation curve depends on the level of active power generation, and therefore, the reactive power can be constrained by $\tan \phi$.

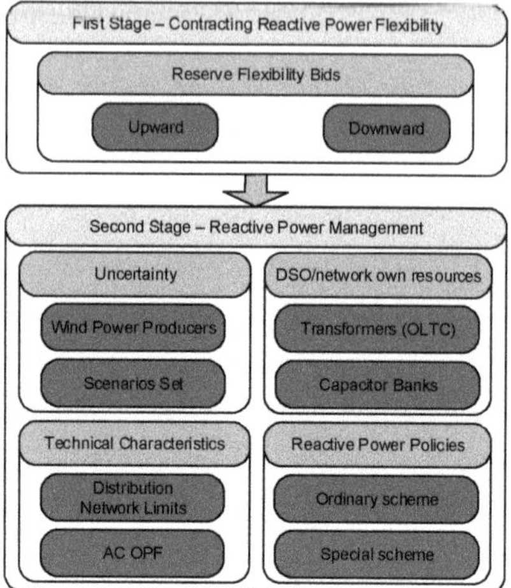

Figure 1. Two-stage optimization model embedded in the sequential AC-optimal power flow (SOPF) tool [15].

In the first stage, the DSO will contract in a day-ahead operating stage, the upward and downward reactive power flexibility that will be used during the operating stage. This represents the maximum reactive power fluctuation of the expected reactive operating point of the DER.

As for the operating stage, the second stage of the model, the DSO will activate the reactive power flexibility needed to overcome the system requirements, respecting the reactive power policy and the technical characteristics of the network (i.e., voltage and thermal limits).

The SOPF tool will then have as its output the amount and type of flexibility the DSO may need to operate the distribution system in expectation, ahead of the operating hour, outputting the upward and downward reactive power flexibility. The upward reactive power flexibility stands for increasing reactive power injection or decreasing reactive power absorbing in the grid, while downward reactive power flexibility stands for the opposite of the upward reactive power flexibility. Note that the reactive power flexibility should be contracted to the distributed generation to maintain the reactive power profile agreed with the TSO while outputting the optimal scheduling of the static equipment (capacitor banks and transformers with on-load tap changer—OLTC) managed by the DSO.

Nevertheless, the SOPF tool is scalable to consider distribution networks with hundreds of nodes, DER, and scenarios at the expense of computational effort. In addition, it does not consider network reconfiguration (which is a common tool for distribution grid management) but can be adapted to include it. Still, this would lead to increased complexity of the tool, hence the computational effort. Therefore, a tradeoff between system complexity and computational effort must be performed to ensure reliable solutions within the DSO operating window for day-ahead simulation.

3. Mathematical Formulation

The proactive reactive power management accounts for the uncertainty factor of renewable energy sources (RES). By utilizing the policies mentioned in Section 3.1, the proposed tool aims to manage the reactive power of the distribution network and to provide reactive power control according the TSO needs.

3.1. Objective Function

The objective Function (2) aims to minimize the operating costs of the DSO to maintain the distribution grid operating within the limits. It includes the costs related to each of the stages, in which the first stage (F^{DA}) comprises the here-and-now decisions and the second-stage (F^{RT}) the wait-and-see decisions.

$$\min \quad F^{DA} + F^{RT} \tag{2}$$

where F^{DA} and F^{RT} are described as in (3) and (4).

$$F^{DA} = \sum_{g=1}^{N_G} \left(C_{DER(g)}^{Q,UP} R_{DER(g)}^{Q,UP} + C_{DER(g)}^{Q,DW} R_{DER(g)}^{Q,DW} \right) + p_{TSO}^{Q,UP} RLX_{TSO}^{Q,UP} + p_{TSO}^{Q,DW} RLX_{TSO}^{Q,DW} \tag{3}$$

$$F^{RT} = \sum_{\omega=1}^{\Omega} \pi_{(\omega)} \left[\begin{array}{l} \sum_{g=1}^{N_G} \left(C_{DER(g)}^{act} \left(r_{DER(g,\omega)}^{Q,UP} - r_{DER(g,\omega)}^{Q,DW} \right) + C_{DER(g)}^{cut} P_{DER(g,\omega)}^{cut} \right) + p_{TSO}^{rlx,act} \left(rlx_{TSO(\omega)}^{Q,UP} - rlx_{TSO(\omega)}^{Q,DW} \right) + p_{TSO(\omega)}^{Extra} rlx_{TSO(\omega)}^{Extra} + \\ \sum_{l=1}^{N_L} \left(C_{(l)}^{DR} P_{(l,\omega)}^{DR} \right) + \sum_{cb=1}^{N_{CB}} \sum_{lv=1}^{N_{levels}} C_{CB(cb)} Z_{CB(cb,\omega,lv)} + \sum_{trf=1}^{N_{TRF}} \sum_{lv=1}^{N_{levels}} C_{TRF(trf)} Z_{TRF(trf,\omega,lv)} \end{array} \right] \tag{4}$$

F^{DA} represents the first-stage decision of contracting reactive power flexibility. Here, DER provides cost inflicted, upward and downward reactive power flexibility. It is mathematically presumed that the TSO request of reactive power may be needed to be relaxed (represented by RLX). This mathematical relaxation proposes the possibility of a certain deviation of the requested $\tan \phi$ value in the upstream connection and is affected by its own penalty. The $\tan \phi$ value is dependent on agreements between the TSO and the DSO.

Concerning F^{RT}, it portrays the real-time operating costs of the distribution network. By the cost of an activation price, generators may change their reactive power operating point. In cases of higher need of flexibility (when the DSO cannot entirely provide the service), a different relaxation is activated through the binary variable rlx^{Extra} allowing the DSO to provide part of the TSO request. By applying an even greater cost, it is possible to curtail the generators active power for relaxing situations where active power is creating problems in the distribution network. Demand response can also be contemplated to decrease the active power consumption, which in turn will reduce the reactive power consumption, under even greater penalties for this relaxation. These alternatives options will ensure that DSO prioritizes DER and consumers over providing the reactive power service to the TSO.

Capacitor banks and the transformers OLTC ability are also considered with a cost related to the lifetime degradation of the equipment by changing the tap set point [16].

3.2. First-Stage Constraints

The first-stage constraints, seen in (5) and (6), represent the DER flexibility for upward and downward reactive power. Similar constraints are applied to the mathematical relaxation of the external supplier flexibility.

$$R_{DER(g)}^{Q,UP,Min} \leq R_{DER(g)}^{Q,UP} \leq R_{DER(g)}^{Q,UP,Max}, \quad \forall g \in \{1,\ldots,N_G\} \tag{5}$$

$$R_{DER(g)}^{Q,DW,Min} \leq R_{DER(g)}^{Q,DW} \leq R_{DER(g)}^{Q,DW,Max}, \quad \forall g \in \{1,\ldots,N_G\} \tag{6}$$

3.3. Second-Stage Constraints

The second-stage constraints refer to the operating stage constraints that are introduced by the uncertainty of RES production. DER active power relates to its operating point for the energy schedule. This value is assumed as fixed by the conditional mean forecast for active power generation. This leads to the active power curtailment in the operating stage to be limited by:

$$P_{DER(g,\omega)}^{cut} \leq P_{DER(g)}^{op} + \Delta P_{DER(g,\omega)}, \quad \forall g \in \{1,\ldots,N_G\}, \forall \omega \in \{1,\ldots,\Omega\} \tag{7}$$

The difference of active power between the realization scenario and the expected forecast in each scenario is represented as ΔP. The active power flowing from the upstream connection (TSO) is limited by the contracted boundaries between the TSO and the DSO and by the capacity of the transformers at the substation interconnection. Active power can be injected/absorbed by the TSO as seen in (8).

$$-P_{TSO(\omega)}^{Max} \leq P_{TSO(\omega)} \leq P_{TSO(\omega)}^{Max}, \quad \forall \omega \in \{1,\ldots,\Omega\} \tag{8}$$

In addition, the second-stage also includes the bounds of the second-stage variables and the non-anticaptivity constraints, given by:

$$r_{DER(g,\omega)}^{Q,UP} \leq R_{DER(g)}^{Q,UP}, \quad \forall g \in \{1,\ldots,N_G\}, \forall \omega \in \{1,\ldots,\Omega\} \tag{9}$$

$$r_{DER(g,\omega)}^{Q,DW} \leq R_{DER(g)}^{Q,DW}, \quad \forall g \in \{1,\ldots,N_G\}, \forall \omega \in \{1,\ldots,\Omega\} \tag{10}$$

Constraints (9) and (10) are also applied to the mathematical relaxation represented through external suppliers.

Each DER has the possibility to provide inductive or capacitive reactive power under the operation limits defined in the Portuguese regulation.

$$-\left(P_{DER(g)}^{op} + \Delta P_{DER(g,\omega)} - P_{DER(g,\omega)}^{cut}\right)\tan\phi \leq Q_{DER(g)}^{op} + r_{DER(g,\omega)}^{Q,UP} - r_{DER(g,\omega)}^{Q,DW} \leq \left(P_{DER(g)}^{op} + \Delta P_{DER(g,\omega)} - P_{DER(g,\omega)}^{cut}\right)\tan\phi,$$
$$\forall g \in \{1,\ldots,N_G\}, \forall \omega \in \{1,\ldots,\Omega\} \tag{11}$$

In (12) and (13), it is represented the upward/downward activation of the mathematical relaxation for the TSO. This relaxation considers a high penalty because the main goal is to provide the service for the TSO request.

$$rlx_{TSO(\omega)}^{Q,UP} \leq RLX_{TSO}^{Q,UP}, \quad \forall \omega \in \{1,\ldots,\Omega\} \tag{12}$$

$$rlx_{TSO(\omega)}^{Q,DW} \leq RLX_{TSO}^{Q,DW}, \quad \forall \omega \in \{1,\ldots,\Omega\} \tag{13}$$

As a last resource to find a solution for congestion and voltage problems, demand response is used by the DSO, being constrained by:

$$P_{L(l,\omega)}^{DR} \leq P_{L(l)}, \quad \forall l \in \{1,\ldots,N_L\}, \forall \omega \in \{1,\ldots,\Omega\} \tag{14}$$

Then, the actual reactive power consumption of consumer *l* is given by:

$$Q_{L(l,\omega)} = \left(P_{L(l)} - P_{L(l,\omega)}^{DR}\right)\tan\phi, \quad \forall l \in \{1,\ldots,N_L\}, \forall \omega \in \{1,\ldots,\Omega\} \tag{15}$$

where *tan* ϕ can be settled at 0.3 as assumed in [17].

Regarding the capacitor banks and transformers with OLTC, these devices are owned by the DSO and located in the substation. This means that the DSO has the knowledge of their characteristics. Capacitor banks are used to provide reactive power being modelled by levels of reactive power as in (16) and (17).

$$Q_{CB(cb,\omega,lv)} = Q_{CB(cb,lv)}^{levels} X_{CB(cb,\omega,lv)}, \quad \forall cb \in \{1,\ldots,N_{CB}\}, \quad \forall \omega \in \{1,\ldots,\Omega\}, \forall lv \in \{1,\ldots,N_{levels}\} \tag{16}$$

$$\sum_{lv=1}^{N_{levels}} X_{CB(cb,\omega,lv)} = 1, \quad \forall cb \in \{1,\ldots,N_{CB}\}, \forall \omega \in \{1,\ldots,\Omega\} \tag{17}$$

The cost of changing the tap of the capacitor banks is multiplied by Z_{CB}, which represents the difference between the tap selection in the present period with the previous one, which is constrained by:

$$X_{CB(cb,\omega,lv)}^{t-1} - X_{CB(cb,\omega,lv)} \leq Z_{CB(cb,\omega,lv)}, \tag{18}$$

$$X_{CB(cb,\omega,lv)} - X^{t-1}_{CB(cb,\omega,lv)} \leq Z_{CB(cb,\omega,lv)}, \quad \forall cb \in \{1,\dots,N_{CB}\}, \forall \omega \in \{1,\dots,\Omega\}, \forall lv \in \{1,\dots,N_{levels}\} \tag{19}$$

The transformers with OLTC constraints for voltage control are modelled as:

$$\Delta V_{TRF(trf,\omega,lv)} = V^{levels}_{TRF(trf,lv)} X_{TRF(trf,\omega,lv)}, \quad \forall \omega \in \{1,\dots,\Omega\}, \forall trf \in \{1,\dots,N_{TRF}\}, \forall lv \in \{1,\dots,N_{levels}\} \tag{20}$$

$$\sum_{lv=1}^{N_{levels}} X_{TRF(trf,\omega,lv)} = 1, \quad \forall \omega \in \{1,\dots,\Omega\}, \forall trf \in \{1,\dots,N_{TRF}\} \tag{21}$$

$$V_{sb(\omega)} = V^{ref}_{sb(\omega)} + \sum_{lv=1}^{N_{levels}} \Delta V_{TRF(trf,\omega,lv)}, \quad \forall \omega \in \{1,\dots,\Omega\}, \forall trf \in \{1,\dots,N_{TRF}\} \tag{22}$$

where ΔV_{TRF} represents the voltage level to be activated in the transformer by the DSO. V^{levels}_{TRF} is a parameter representative of all possible taps of the transformer, and X_{TRF} is the binary variable for selection of a unique tap level. V^{ref}_{sb} is the reference of voltage magnitude at the substation before the use of OLTC ability by the transformer, while the final voltage value at the substation is denoted by V_{sb}. In addition, the cost for changing the tap of the transformer is included in the objective function (5), where Z_{TRF} is the linearization of the absolute function, as the capacitor banks. Thus, the constraints are:

$$X^{t-1}_{TRF(trf,\omega,lv)} - X_{TRF(trf,\omega,lv)} \leq Z_{TRF(trf,\omega,lv)}, \tag{23}$$

$$X_{TRF(trf,\omega,lv)} - X^{t-1}_{TRF(trf,\omega,lv)} \leq Z_{TRF(trf,\omega,lv)}, \quad \forall trf \in \{1,\dots,N_{TRF}\}, \forall \omega \in \{1,\dots,\Omega\}, \forall lv \in \{1,\dots,N_{levels}\} \tag{24}$$

Moreover, an AC-OPF is used to model the power flow in the distribution network. Therefore, the active power balance in each bus is modelled as:

$$\sum_{g=1}^{N_G} \left(P^{op,i}_{DER(g)} + \Delta P^i_{DER(g,\omega)} - P^{cut}_{DER(g,\omega)} \right) + P^i_{TSO} + \sum_{l=1}^{N_L} \left(P^{DR,i}_{L(l,\omega)} - P^i_{L(l)} \right) = G_{ii}V^2_{i(\omega)} + V_{i(\omega)} \sum_{j \in TL^i} V_{j(\omega)} \left(G_{ij}\cos\theta_{ij(\omega)} + B_{ij}\sin\theta_{ij(\omega)} \right),$$
$$\forall i \in \{1,\dots,N_{Bus}\}, \forall \omega \in \{1,\dots,\Omega\}, \theta_{ij(\omega)} = \theta_{i(\omega)} - \theta_{j(\omega)} \tag{25}$$

Additionally, the reactive power balance is given by:

$$\sum_{g=1}^{N_G} \left(Q^{op,i}_{DER(g,\omega)} + r^{Q,UP,i}_{DER(g,\omega)} - r^{Q,DW,i}_{DER(g,\omega)} \right) - \sum_{l=1}^{N_L} Q^i_{L(l,s)} + \sum_{cb=1}^{N_{CB}} \sum_{lv=1}^{N_{levels}} Q^i_{CB(cb,\omega,lv)} + Q^{op,i}_{TSO(\omega)} + rlx^{Q,UP,i}_{TSO(\omega)} - rlx^{Q,DW,i}_{TSO(\omega)} + rlx^{Extra,i}_{TSO(\omega)} =$$
$$V_{i(\omega)} \sum_{j \in TL^i} V_{j(\omega)} \left(G_{ij}\sin\theta_{ij(\omega)} - B_{ij}\cos\theta_{ij(\omega)} \right) - B_{ii}V^2_{i(\omega)}, \quad \forall i \in \{1,\dots,N_{Bus}\}, \forall \omega \in \{1,\dots,\Omega\}, \theta_{ij(\omega)} = \theta_{i(\omega)} - \theta_{j(\omega)} \tag{26}$$

There is also the consideration that the energy flowing through the distribution lines has a thermal limit that should not be exceeded, being limited as in (27) and (28).

$$\left| \overline{V_{i(\omega)}} \left[\overline{y_{ij}V_{ij(\omega)}} + \overline{y_{sh(i)}V_{i(\omega)}} \right]^* \right| \leq S^{Max}_{TL}, \overline{V_{ij(\omega)}} = \overline{V_{i(\omega)}} - \overline{V_{j(\omega)}}, \quad \forall i,j \in \{1,\dots,N_{Bus}\}, \forall \omega \in \{1,\dots,\Omega\}, i \neq j \tag{27}$$

$$\left| \overline{V_{j(\omega)}} \left[\overline{y_{ij}V_{ji(\omega)}} + \overline{y_{sh(j)}V_{j(\omega)}} \right]^* \right| \leq S^{Max}_{TL}, \overline{V_{ji(\omega)}} = \overline{V_{j(\omega)}} - \overline{V_{i(\omega)}}, \quad \forall i,j \in \{1,\dots,N_{Bus}\}, \forall \omega \in \{1,\dots,\Omega\}, i \neq j \tag{28}$$

Voltage magnitude must stay between the limits established by the DSO, assuming the slack bus voltage magnitude as fixed.

$$V^i_{Min} \leq V_{i(\omega)} \leq V^i_{Max}, \quad \forall \omega \in \{1,\dots,\Omega\} \tag{29}$$

4. Case Study

This section presents the case study used to apply and test the model developed for the Portuguese reactive management policies. The simulation has been carried out with MATLAB and GAMS tools.

4.1. 37-Bus Distribution System

The present case study is based on a 37-bus distribution network (originally presented in [18]) that was adapted to support five DER in the form of three combined heat and power (CHPs) and two wind turbines, as one can see in Figure 2.

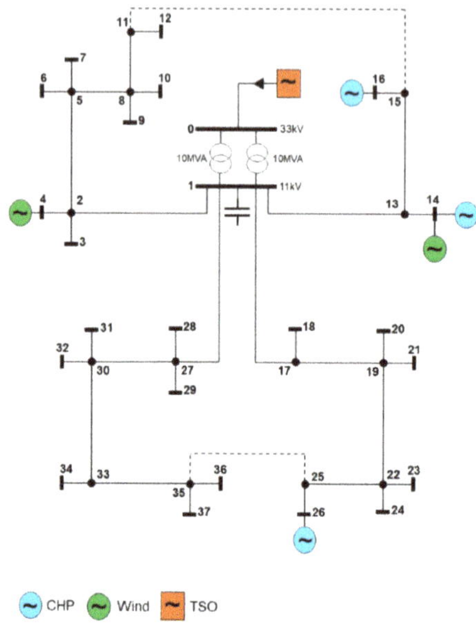

Figure 2. 37-Bus distribution network (adapted from [18]).

The distribution network is connected to a high voltage network through two power transformers of 10 MVA each. It also possesses 22 consumption points that represent 1908 consumers (1850 residential consumers, two industrial consumers, 50 commercial stores, and six service buildings) [18], with the consumption characteristics and profile being adopted from [19]. The total active and reactive power consumption in the network is summarized in Table 2.

There are two transformers and two capacitor banks being considered in the network. More precisely, the transformers have OLTC ability with a maximum voltage deviation of 0.1 p.u. In addition, the capacitor banks also have steps with a total capacity of reactive power production of 5.4 MVAr (one capacitor bank with 4.5MVAr and the other with 0.9 MVAr). The cost reflecting the use of the transformers and capacitor banks (with the OLTC ability that reduces the equipment lifetime) is determined by [16]. It is assumed that both types of equipment are owned and managed by the DSO.

The network is composed of different DER. More precisely, three CHP units and two wind turbines which can provide reactive power flexibility, accordingly to their technical limits.

It is assumed that the DER active power generation should be fully absorbed by the network, following the standard regulation. Therefore, Table 3 shows the generic characteristics of the DER, including the expected operating point (e.g., wind power forecast).

The DER characteristics, as well as the cost for upward and downward reactive power flexibility, are given in Table 4.

Table 2. Active and reactive power consumption characteristics.

Load.Bus		Active Power Consumption (kW)			Reactive Power Consumption (kVAr)		
		Min	Mean	Max	Min	Mean	Max
1	3	373.2	677.9	1190.5	112.0	203.4	357.2
2	4	206.1	591.2	1015.6	61.8	177.4	304.7
3	6	88.4	599.0	1029.8	26.5	179.7	308.9
4	7	394.7	716.9	1259.1	118.4	215.1	377.7
5	9	539.0	761.8	1089.0	161.7	228.5	326.7
6	10	298.7	636.6	1040.9	89.6	191.0	312.3
7	12	323.0	586.5	1030.1	96.9	176.0	309.0
8	14	387.0	1110.4	1907.4	116.1	325.4	567.1
9	16	745.6	1589.1	2598.3	223.7	425.6	779.5
10	18	509.7	720.3	1029.8	152.9	169.7	308.9
11	20	88.4	599.0	1029.8	26.5	152.1	308.9
12	21	373.2	677.9	1190.5	112.0	190.9	357.2
13	23	365.1	778.1	1272.3	109.5	208.4	381.7
14	24	539.0	761.8	1089.0	161.7	179.5	326.7
15	26	323.0	586.5	1030.1	96.9	165.1	309.0
16	28	178.3	511.6	878.8	53.5	149.9	261.3
17	29	74.4	503.8	866.2	22.3	128.0	259.9
18	31	314.0	570.2	1001.4	94.2	160.5	300.4
19	32	290.4	618.9	1011.9	87.1	165.8	303.6
20	34	93.5	633.4	1089.0	28.1	160.9	326.7
21	36	217.9	625.3	1074.1	65.4	183.2	319.3
22	37	323.0	586.5	1030.1	96.9	165.1	309.0

Table 3. General characteristics and operating point for distributed energy resources (DER).

DER	Number of Units	Total Installed Power	Operating Point P^{op} (MW)		
			Min	Mean	Max
CHP	3	2.5 (MVA)	1.0	1.15	1.5
Wind	2	20 (MVA)	11.31	14.01	15.34
Transmission system operator (TSO)	1	20 (MVA)	-	-	-

Table 4. DER reactive power costs.

DER	Upward Cost C^{up} (m.u./kVAr)			Downward Cost C^{dw} (m.u./kVAr)		
	Min	Mean	Max	Min	Mean	Max
CHP	0.02	0.04	0.06	0.02	0.04	0.06
Wind	0.02	0.025	0.03	0.02	0.025	0.03
TSO	1	1	1	1	1	1

RESs are modelled through stochastic variables. Thus, upward and downward reactive power flexibility is constrained by their technical limits. In [20,21] can be found the scenarios used to model the uncertainty of wind power forecast. A set of 10 scenarios were extracted for each wind generator. In this case study, the standard reactive power policy of the DER is subjected to the Portuguese regulation, following Table 1.

Regarding the upstream connection, it must be established the *tan* ϕ agreed between the TSO and the DSO for the substation of interconnection. In this case, it has been considered that the *tan* ϕ varies throughout the day according to the regulation established in [12]. More precisely, the TSO must provide a *tan* ϕ of 0 with +/−5% of deviation between 22:00 and the 07:00. In the remaining period, the *tan* ϕ is expected to be 0.3 with +/−5% deviation. Note that the *tan* ϕ can vary from these values, taking into account specific agreement between the DSO and the TSO for a specific substation of interconnection. The case study was constructed assuming that active power from DER can be greater or less than the load on the network. Thus, the TSO can either inject or absorb active power depending on the realization of wind generation over time.

4.2. Results

The tool will attempt, by using capacitor banks, transformers with OLTC and DER, to contract optimal reactive power flexibility and with this, meet the desired reactive power profile defined by the DSO.

It is important to note that active power production in the DERs is fixed according to the previous forecast of the energy dispatched for each hour. Wind power plants have a forecast point determined for the next 24 h along with 10 possible hourly scenarios, as in Figure 3.

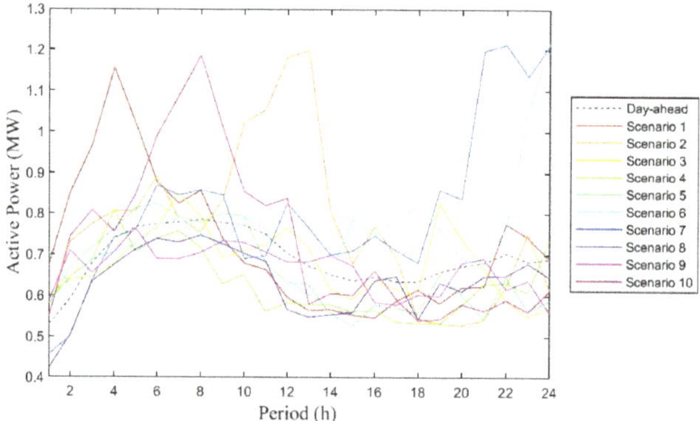

Figure 3. Point forecast and scenarios of the wind power plant 1 throughout 24 h.

Using wind active power as an input and considering the flexibility limits, the expected reactive power production operation is determined by the tool for each operation hour point. The *tan* ϕ of both wind power plants is kept within the +/−5% range in every scenario. Figure 4 shows the evolution of *tan* ϕ overtime for the wind power plant 1.

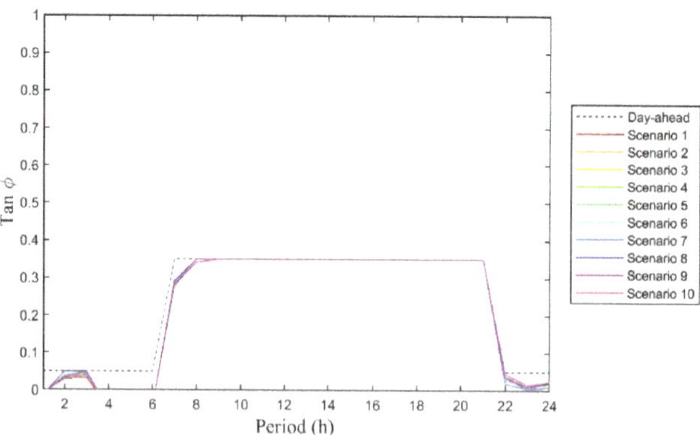

Figure 4. Forecasted and realized tan ϕ of the wind power plant 1 throughout 24 h.

Figure 5 depicts the tan ϕ profile at the substation. As was predicted, the *tan* ϕ values are as close as possible to 0.3 between 7–22 h and a value of 0 for *tan* ϕ for the other hours, for every scenario, to guaranty that no penalties to be applied to the DSO. DERs reactive power generation, the capacitor banks and the transformers with OLTC ability have an important role in securing this result.

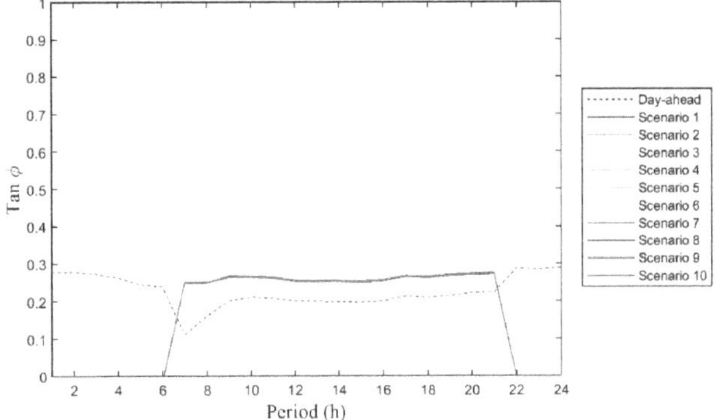

Figure 5. Forecasted and realized *tan* ϕ at the main substation throughout 24 h.

Figure 6 shows the reactive power production of capacitor bank 1. The level of reactive power varies according to the DSO needs. For periods between 22:00 and 07:00, the level of reactive power production is high, reaching the maximum production in some periods. During the day, the reactive power production of the capacitor bank comes to zero, since the TSO is injecting a significant amount of reactive power which is sufficient to support the system. It is also worth mentioning that the step position does not change more than four times per day, which reduces equipment degradation over time.

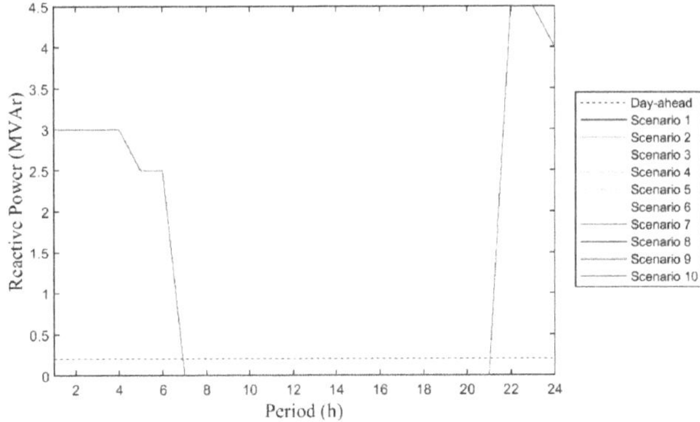

Figure 6. Reactive power production of capacitor bank 1 throughout 24 h.

Capacitor bank 2 follows the same behavior as capacitor bank 1, as can been seen in Figure 7. In fact, as capacitor bank 2 is much smaller than capacitor bank 1, the capacitor bank 2 is often used to complement the reactive power between steps of the capacitor bank 1.

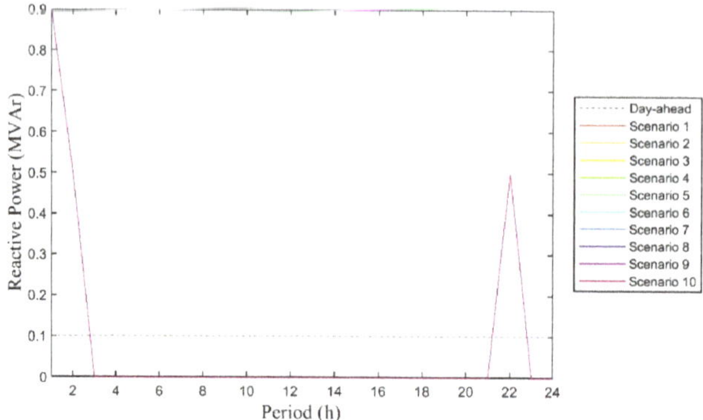

Figure 7. Reactive power production of capacitor bank 2 throughout 24 h.

As with the capacitor banks, OLTC tap changes are reduced, even maintaining the same position throughout the 24 h. It is noteworthy that as the optimization considers the day-ahead forecast point, it leads to the modification of the OLTC tap hourly positions (Figure 8). Note that both transformers present the same behavior as presented in Figure 8.

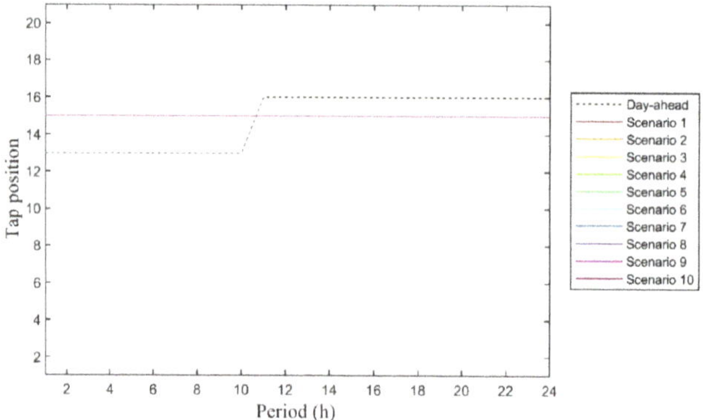

Figure 8. Transformer 1 with OLTC ability throughout 24 h.

5. Conclusions

This work proposes a new tool to be used by the DSO in reactive power management, exploiting DER flexibility. It explores the use of a two-stage stochastic model that manages the uncertainty of wind power producers. With this tool, TSO reactive power requirements can be provided by contracting the service to the DSO, which may be an alternative to investments in reactive power control equipment in the transmission network. Simulations were done for a 37-bus distribution network, whose results demonstrate the feasibility of the proposed tool. The selection of the DER that could provide reactive power flexibility, under the different operation conditions introduced, was proven and the service was provided to the TSO.

Author Contributions: Conceptualization, T.S. (Tiago Soares), L.C., H.M., T.S. (Tiago Simão) and M.L; Methodology, T.S. (Tiago Soares), L.C. and H.M.; Software, T.A. and T.S. (Tiago Soares); Validation, T.A., T.S. (Tiago Soares), L.C. H.M., T.S. (Tiago Simão) and M.L; Visualization, T.A. and T.S. (Tiago Soares); Writing—original draft, T.A., T.S. (Tiago Soares), L.C. and H.M., Writing—review & editing, T.A., T.S. (Tiago Soares), L.C., H.M, T.S. (Tiago Simão) and M.L.

Funding: This work was financed by the European Union's Horizon 2020 through the EU framework Program for Research and Innovation 2014–2020, within the EU-TDX-ASSIST project under the agreement No. 774500.

Energies **2019**, *12*, 4028

This work was also supported by national funds through Fundação para a Ciência e Tecnologia with reference UID/CEC/50021/2019.

Conflicts of Interest: The authors declare no conflict of interest. The funders had no role in the design of the study; in the collection, analyses, or interpretation of data; in the writing of the manuscript, or in the decision to publish the results.

Nomenclature

Parameters

ΔP	Power deviation in each scenario
B	Imaginary part in admittance matrix
C	Cost
G	Real part in admittance matrix
N	Number of unit resources
p	Penalty for external supplier's flexibility
$\overline{y}$	Series admittance of line that connects two buses
$\overline{y}_{sh}$	Shunt admittance of line that connects two buses

Variables

θ	Voltage angle
P	Active power
Q	Reactive power
r	Reactive power flexibility used in the operating stage
rlx	Reactive power relaxation in the operating stage
R	Reactive power flexibility contracted at day-ahead stage
RLX	Reactive power relaxation at day-ahead stage
S	Apparent power
V	Voltage magnitude
$\overline{V}$	Voltage in polar form
V_{sb}	Voltage at slack bus
ΔV	Voltage level activated by the DSO in the transformer
X	Binary variable
Z	Auxiliary variable for absolute function linearization

Subscripts

ω	Index of scenarios
cb	Index of capacitor bank units
CB	Capacitor bank abbreviation
g	Index of generators units
i, j	Bus index
l	Index of load consumers
L	Load consumers abbreviation
lv	Index of levels (tap changing) for capacitor banks and transformers
TSO	Transmission system operator
t	Time index
trf	Index of transformer units
TRF	Transformer abbreviation

Superscripts

act	Activation cost of resources in real-time stage
cut	Generation curtailment
Max	Maximum limit
Min	Minimum limit
op	Operating point of the power resource
Q, DW	Downward reactive power flexibility
Q, UP	Upward reactive power flexibility
DR	Demand response of consumer l

References

1. Pérez-Arriaga, I.J.; Ruester, S.; Schwenen, S.; Battle, C.; Glachant, J.-M. From distribution networks to smart distribution systems: Rethinking the regulation of European electricity DSOs. *Util. Policy* **2014**, *31*, 229–237.
2. Eurelectric Active Distribution System Management: A Key Tool for the Smooth Integration of Distributed Generation. Available online: https://www.eurelectric.org/media/1781/asm_full_report_discussion_paper_final-2013-030-0117-01-e.pdf (accessed on 22 October 2019).
3. Soares, T.; Bessa, R.J. Proactive management of distribution grids with chance-constrained linearized AC OPF. *Int. J. Electr. Power Energy Syst.* **2019**, *109*, 332–342. [CrossRef]
4. Morais, H. TDX-ASSIST: Agreed Models, Use Case List, and Use Case Description in UML, D1.2. Available online: https://ec.europa.eu/research/participants/documents/downloadPublic?documentIds=080166e5bfb19910&appId=PPGMS (accessed on 22 October 2019).
5. Gayatri, M.T.L.; Parimi, A.M.; Pavan Kumar, A.V. A review of reactive power compensation techniques in microgrids. *Renew. Sustain. Energy Rev.* **2018**, *81*, 1030–1036. [CrossRef]
6. Parida, S.K.; Srivastava, S.C.; Singh, S.N. A review on reactive power management in electricity markets. *Int. J. Energy Sect. Manag.* **2011**, *5*, 201–214. [CrossRef]
7. Mohseni-Bonab, S.M.; Rabiee, A. Optimal reactive power dispatch: A review, and a new stochastic voltage stability constrained multi-objective model at the presence of uncertain wind power generation. *IET Generation, Transm. Distrib.* **2017**, *11*, 815–829. [CrossRef]
8. Rabiee, A.; Soroudi, A.; Mohammadi-Ivatloo, B.; Parniani, M. Corrective voltage control scheme considering demand response and stochastic wind power. *IEEE Trans. Power Syst.* **2014**, *29*, 2965–2973. [CrossRef]
9. Wang, L.; Wang, X.; Jiang, C.; Yin, S.; Yang, M. Dynamic coordinated active-reactive power optimization for active distribution network with energy storage systems. *Appl. Sci.* **2019**, *9*, 1129. [CrossRef]
10. Kang, S.J.; Kim, J.; Park, J.W.; Baek, S.M. Reactive power management based on voltage sensitivity analysis of distribution system with high penetration of renewable energies. *Energies* **2019**, *12*, 1493. [CrossRef]
11. Stock, D.S.; Sala, F.; Berizzi, A.; Hofmann, L. Optimal control of wind farms for coordinated TSO-DSO reactive power management. *Energies* **2018**, *11*, 173. [CrossRef]
12. ERSE Regulamento da Rede de distribuição—Portaria 596-2010. Available online: http://www.erse.pt/pt/electricidade/regulamentos/redededistribuicao/Documents/Portaria596-2010RRTRRD.pdf (accessed on 22 October 2019).
13. ERSE Despacho no3/2010. Available online: http://www.erse.pt/pt/consultaspublicas/consultas/Documents/31_4/Despacho_3_2010.pdf (accessed on 22 October 2019).
14. ERSE Informaçáo Sobre Faturaçáo de Energia Reativa: Princípios e Boas Práticas. Available online: http://www.erse.pt/pt/electricidade/regulamentos/relacoescomerciais/Documents/Recomendaç~oes/Recomendaç~aodeenergiareactiva.pdf (accessed on 22 October 2019).
15. Simão, T.; Gama, P.; Louro, M.; Carvalho, L.; Gloria, G.; Pestana, R.; Reis, F.; Soares, T.; Silva, J. TDX-ASSIST: Beyond state of art in TSO-DSO interoperability—The Portuguese demonstrator. In Proceedings of the 25th International Conference on Electricity Distribution, Madrid, Spain, 3–6 June 2019; pp. 3–6.
16. Zhang, Y.J.; Ren, Z. Optimal reactive power dispatch considering costs of adjusting the control devices. *IEEE Trans. Power Syst.* **2005**, *20*, 1349–1356. [CrossRef]
17. ERSE Despacho no 7253/2010. Available online: http://www.erse.pt/pt/legislacao/Legislacao/Attachments/1411/Despacho7253_2010.pdf (accessed on 22 October 2019).
18. Allan, R.N.; Billinton, R.; Sjarief, I.; Goel, L.; So, K.S. A reliability test system for educational purposes—Basic distribution-system data and results. *IEEE Trans. Power Syst.* **1991**, *6*, 813–820. [CrossRef]
19. Morais, H.; Sousa, T.; Soares, J.; Faria, P.; Vale, Z. Distributed energy resources management using plug-in hybrid electric vehicles as a fuel-shifting demand response resource. *Energy Convers. Manag.* **2015**, *97*, 78–93. [CrossRef]
20. Pinson, P. Wind energy: Forecasting challenges for its operational management. *Stat. Sci.* **2013**, *28*, 564–585. [CrossRef]
21. Bukhsh, W.A.; Zhang, C.; Pinson, P. Data for Stochastic Multiperiod OPF Problems. Available online: https://sites.google.com/site/datasmopf/ (accessed on 22 October 2019).

Article

Using Agent-Based Customer Modeling for the Evaluation of EV Charging Systems

Tobias Rodemann [1,*], **Tom Eckhardt** [2], **René Unger** [2] and **Torsten Schwan** [2]

[1]   Honda Research Institute Europe GmbH, 63073 Offenbach am Main, Germany
[2]   EA Systems Dresden GmbH, 01187 Dresden, Germany,
*   Correspondence: tobias.rodemann@honda-ri.de; Tel.: +49-69-89011-732

Received: 19 June 2019; Accepted: 19 July 2019; Published: 25 July 2019

Abstract: The development of efficient electric vehicle (EV) charging infrastructure requires a modeling of customer behavior at an appropriate level of detail. Since only limited information about real customers is available, most simulation approaches employ a stochastic approach by combining known or estimated customer features with random variations. A typical example is to model EV charging customers by an arrival and a targeted departure time, plus the requested amount of energy or increased state of charge (SoC), where values are drawn from normal (Gaussian) distributions with mean and variance values derived from user studies of obviously limited sample size. In this work, we compare this basic approach with a more detailed customer model employing a multi-agent simulation (MAS) framework in order to investigate how a customer behavior that responds to external factors (like weather) or historical data (like satisfaction in past charging sessions) impacts the essential key performance indicators of the charging system. Our findings show that small changes in the way customers are modeled can lead to quantitative and qualitative differences in the simulated performance of EV charging systems.

Keywords: EV charging; multi-agent system; digital twin; customer satisfaction indicator

1. Introduction

With an increasing number of electric vehicles (EVs), an improved charging infrastructure is required in order to deliver the necessary energy for mobility while at the same time reducing the costs of operating said infrastructure. For the development of such charging systems, developers have to optimize a number of system parameters like the number, location, and type of charging stations (CSs). A key factor in this analysis is an accurate representation of customers and their charging requirements. In the majority of prior literature (for example [1,2], see also [3,4]), customers were modeled passively via their charging process with an arrival and departure time of the car plus an energy requirement. These values are typically described as a Gaussian probability distribution with a defined mean and variance. An example customer could be described as arriving at the charging station at 9:00 h $\pm$ 30 min, leaving at 17:30 h $\pm$ 10 min, and requesting 8 ± 0.5 kWh energy. For example, the work in [5] studied the psychological aspects concerning the charging behavior of electric vehicle drivers. During a six-month study with 79 EV users, charging data were collected. Typically, users drove almost 40 km per day and recharged three times weekly, with a high remaining battery energy level at the time of charging.

This approach has a number of benefits. It requires limited information about customers, can easily be scaled, has a low computational expense (data tables can be computed upfront) and is easy to investigate. The drawback is that with increasing numbers of customers, normal distributed variations level out, thus only covering average conditions, potentially overlooking rare, but serious extreme conditions. In addition, as we will show in this work, there are individual effects that cannot be accurately modeled with the conventional approach. In order to show this, we extend the standard approach for modeling EV customers in two separate steps. In a first step, we introduce an EV sensitivity to the current weather conditions. We assume that under very cold or hot conditions, EVs have an increased energy demand, requiring more energy than usual (e.g., due to air conditioning, heating, or battery properties). The quality of the charging process is evaluated using a customer satisfaction indicator (CSI) that compares the result of the charging process (the reached SoC) with the customer requirements. We hypothesize that depending on the increase in energy demand, the mean CSI degrades on certain days. We call this the Temp scenario. The second studied extension (CSI sensitive) is that we assume that customers have a memory of past charging sessions, in the form of CSI values. If CSI values degrade below a threshold, customers will stop using the charging station, and the system will be used by fewer EVs.

The study system used in this work is a charging infrastructure of a medium-sized company. We assume that some of the employees are using EVs to commute and use the low-cost electricity at the company for most of their charging demands. In order to minimize CO_2 emissions, the company uses a large-scale PV system to generate inexpensive, CO_2 -free energy for the charging station, supported by a low level of grid power. The drawback of the approach is that during days without sunshine, only a small level of extra charging power from the grid is available. This approach minimizes CO_2 emissions and energy costs (especially peak load costs), but is sometimes not able to satisfy all customer (EV user) demands.

The practical implementation of customer models that respond to external conditions and historical data is done via a multi-agent system (MAS) approach, where each customer, electric vehicle (EV), and charging station (CS) is represented as a computer program that can respond to current and past conditions. Using this approach, it is possible to replicate the conventional approach with normal randomly-distributed requests, but also to investigate other more realistic scenarios. In our study, we employ the NetLOGO framework.

Using the baseline and the two extended scenarios, we show how the characterized modeling framework can lead to new perspectives and insights on the more complex interactions between charging infrastructure systems and their customers.

1.1. Related Work

Multi-agent systems have been used intensively in recent year in the energy sector. In [4], the impact of the variation of charging prices on the behavior of electric vehicle users was investigated, comparing benefits when using variable and fixed charging prices. With the goal of generating a realistic population of customers, an EV usage simulator was developed that models the effects of city size. Offering variable prices showed good results and could be a promising approach where EV users have different prices when deciding the location of their next charging spot.

Xydas et al. [6] built a MAS for charging controllers and EVs. The objective of the controller was to operate the CS in order to perform peak shaving, valley filling, maximizing income, and for the EV to minimize costs in two different scenarios, one where the agent is reacting to variations of electricity prices, and one for charging processes irrespective of price signals. The aim of the research presented in [7] was to measure user satisfaction and their adaptability to new vehicle technologies (specifically, full electric vehicles). The work in [8] presented a simulation framework for electric vehicles in terms of energy consumption. In [9], vehicle behavior was simulated by an agent-based transportation simulation tool with detailed routes on a map of Zurich for the simulation of vehicle behavior with realistic vehicle energy consumption profiles. The approach uses multi-agent transportation simulation (MATSim). Sweda et al. [10] looked at the strategic investment in new charging infrastructure considering private EV usage and driving patterns. The study was done using an MAS to simulate user behaviors.

1.2. Multi-Agent Simulation Systems

In the literature, different MAS frameworks have been employed. In [11], the authors used an MAS for investigating the issue of avoiding grid congestion in scenarios with a large number of EVs. The simulation was based on the JACK framework in conjunction with MATLAB/Simulink.

In [12], the authors briefly summarized multi-agent simulation tools and specifically evaluated MASON and NetLOGO via a test implementation for human modeling (We briefly cite one of their main conclusions: "NetLOGO has proved its reputation as an ABS platform where the simulation models can be implemented quickly and straightforwardly. ... Had we simply looked for a handy standalone agent-based simulation tool for a limited number of agents, NetLOGO easily could have been our choice..."). NetLOGO was also employed in [13] to simulate human behavior.

Another comparison study on MAS tools as [14]. They also came to the conclusion that NetLOGO was surprisingly efficient for the task ("Perhaps because NetLOGO is clearly designed for one type of model (Section 3.1) and uses a simplified language, scientists tend to assume it is too limited for serious ABMs. We originally intended to exclude NetLOGO as too limited for full treatment in this paper, but found we could implement all our test models (Section 2.1) in NetLOGO, with far less effort than for other platforms.").

2. Methods

In this section, we outline the relevant components of the tool chain, the first being the simulation environment for the energy system, which employs a commercial tool (SimulationX [15]) based on the Modelica [16] standard. The simulation communicates via a functional mockup interface (FMI) [17] with two external programs, the charging controller implemented in Python and the multi-agent simulation tool NetLOGO.

2.1. Modelica Simulation Environment

SimulationX is a multi-physics program package for modeling, calculation, simulation, optimization, and reliability analysis of technical components and systems. It is based on the object-oriented programming language Modelica and supports the standardized interface FMI for communication between SimulationX and external programs. We also employed the GreenCity extension library that contains components for the simulation of a complete energy system with a focus on renewables. The main benefit of this simulation approach is that the engineering know-how implemented in the library can easily be accessed so that developers can directly focus on their key research areas (for example, controller development). Simulations of a complete year run on time-scales of seconds to minutes and can be parallelized for parameter optimization [18].

2.2. Functional Mockup Interface

FMI is a software standard to facilitate co-simulation approaches in complex simulation tasks involving several domains. The basic idea of FMI within a co-simulation is to integrate two or more simulation tools into a common simulation environment. The data exchange takes place at defined communication times. In the time between the communication points, the tools simulate their models independently of each other. A component that has implemented the FMI standard is called an FMU. The FMU is a compressed archive file (*.fmu) containing a model description and the corresponding libraries (*.dll). For more information, see [17,19] or www.fmi.org.

2.3. Charging Controller

The charging controller is a tool written in Python. It contains all the parameters of the charging processes and is responsible for controlling the charging processes according to predefined algorithms The controller thus receives information on the physical conditions of the charging stations, the physical parameters of the arriving vehicles, and information on the usage of the charging station.

3. Multi-Agent Simulator

This section briefly describes the basics of multi-agent simulation (MAS). NetLOGO was chosen as the tool for the implementation of the MAS, which will be explained further below. In Section 3.3, the most important elements of the programmed MAS are presented concisely.

3.1. Basic MAS Philosophy

Multi-agent systems are comprised of autonomously-operating agents and the environment in which they are located. The goal of the systems is to solve complex problems through specializations and division of labor. Well-known examples from biology are bee swarms and ant colonies. In these collectives, a clear differentiation takes place between the individual member groups. In the MAS, an agent can be defined as a "[...] separable and definable software unit, that interacts flexibly and autonomously with its environment and other agents, so as to achieve the predefined goals of the user or developer, ultimately adhering to his interests." (translated from [20], see also [21]).

3.2. NetLOGO

NetLOGO is an independent programming language, which includes a programming environment that is, however, fundamentally based on JAVA. This tool offers the possibility to be accessed by superordinate JAVA programs. We integrated it into co-simulation by creating a wrapper software which accesses the NetLOGO API and wraps it into an FMU (functional mock-up unit). A main disadvantage is the limited visual representation of the simulation. Methods need to be programmed independently; however, the programming language can be quickly learned by novices. This tool has gained popularity in the fields of teaching and research. NetLOGO contains a large library of complete exemplary models that can be used and expanded (the download of the software and further information can be found on the following website: http://ccl.northwestern.edu/netlogo/index.shtml).

3.3. Key Features of Our Implementation

The model contains three different agent types and a model environment. All agent types differentiate themselves in their fundamental characteristics, features, actions, and their communication. Of each agent type, multiple individual agents can exist. In this specific model example, 30 agents of the type "e-vehicle", 21 agents of the type "charging point", and 30 of the "driver" agent type were implemented. The model additionally contains the "environment". This encapsulates all actions and parameters that are agent independent or not agent specific (e.g., time of day).

The agent "driver", largely simplified, represents the human driver. He/she uses his/her vehicle and the appropriate charging infrastructure to accomplish his/her daily routine. The daily routine is based on statistical and normal distributed random numbers. The emphasis is on getting from A to B with sufficient battery capacity. This means that a vehicle does not charge at every opportunity, but only when the driver considers it necessary. In concrete terms, this is implemented in that a charging decision is made when the SoC falls below an individual threshold value SoC_{Limit}. This threshold value depends on where the vehicle is located, meaning whether it charges at the company (SoC_{Limit} = 75%) or at an external charging infrastructure (SoC_{Limit} = 20%). The agent "e-vehicle" (EV) is a digital representation of a real electric vehicle. These agents have a reduced pool of parameters, which are necessary for the simulation. These include information on charging infrastructure, consumption, and charging status. The vehicles have rudimentary functions such as "driving", "parking", and "loading". Within these methods, the vehicle's batteries are emptied or charged accordingly. Furthermore, simulation-relevant information such as ambient temperature or charging power is transmitted to the agent via the FMU. The agents of the type "charging point" (CP) each represent a single charging point in the real charging infrastructure. These agents have individual charging parameters, allowing the "driver" agents to charge the "e-vehicle" agents if the parameters match between EV and CP. Particularly important is the role of the CP agent as the model interface enabling the saving and export of all the charging process information from the model.

4. Scenarios

In this work, we shall compare three different scenarios, based on different customer behavior. The first scenario represents conventional static usage profiles based on normal distributed charging demands. This type of customer modeling could be implemented without an MAS approach; however, we decided to also employ our MAS framework for better comparability. The second scenario is a minor extension of the basic scenario where the energy demand on hot or cold days is increased. We assume that for technical reasons, the energy demand of the EV increases due to the battery chemistry or additional A/C and heating demand. In the third scenario, the customer shall stop using our charging infrastructure if past charging processes were not to the customer's liking. This is modeled by a customer satisfaction indicator that is computed based on the difference of targeted and actual reached SoC for past charging sessions. The assumption is that if the charging infrastructure repeatedly failed to charge the EV to the requested level, the customer would switch to a different charging operator. More details on the three scenarios are given below.

4.1. Building Scenario

Our test system was a medium-sized company with $N_E = 30$ EVs, $N_{CS} = 21$ charging stations, and a photovoltaic (PV) system with a peak power of $P_{PV} = 490$ kW (for the output of the PV system over a year, see Figure 1). The rest of the facility was not modeled in detail. Charging stations and EVs have different maximum charging power between 10 kW and 150 kW.

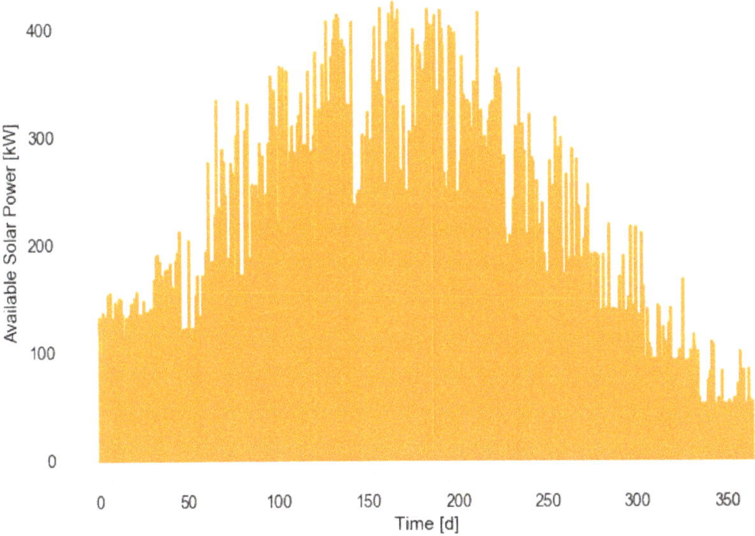

Figure 1. PV system output during the year.

4.2. Charging Controller

We used a simple charging controller in this project. The controller first computes the currently available charging power: P_a:

$$P_a(t) = P_{PV}(t) + P_{Base} \quad ,$$

(1)

where P_{PV} is the current power production from the PV system and P_{Base} additional power taken from the grid. In order to minimize peak loads and to reduce costs and CO_2 emissions, one would try to minimize P_{Base}. Here, we have chosen $P_{Base} = 30$ kW, but also consider two extreme scenarios with $P_{Base} = 0$ kW and $P_{Base} = 1$ MW (MAX scenario). These values were chosen to cover three interesting conditions: one where only PV power is available ($P_{Base} = 0$ kW), which means 100% green energy, the additional availability of on average 1 kW per EV ($P_{Base} = 30$ kW), and finally, a setting for virtually unlimited grid power ($P_{Base} = 1$ MW). These values both span a wide range, but also appear realistic under different ecological or economic constraints.

As an example, we note that at our facility, peak costs of approximately 100 Euros/kW occur, so that P_{Base} should be kept as low as possible.

In the next step, the available power P_a was evenly distributed between all EVs until they reached their target SoC. If an EV was provided with more charging power than what was technically feasible, this surplus power was returned to the system and made available for vehicles with higher potential charging power. Only when all EVs reached their target SoC values, then all vehicles were charged from the target SoC to 100%, again evenly distributed.

Note that our simple controller did not consider the current SoC in the EV to prioritize EVs with low SoC levels, since this would increase the complexity of the controller and also raise issues of fairness. We also neglected the impact of the charging strategy on the battery state of health (SOH). While the simulator performed a basic SOH modeling depending on cyclic and calendaric aging, these effects were not considered in the controller and our analysis.

4.3. Customer Satisfaction Indicator

In contrast to prior work [22], we only employed a single function for computing the CSI value of a charging process. Based on the desired SoC given by the user (SoC_d), the SoC at the arrival (SoC_{ini}), and the achieved SoC at the customer's departure (SoC_a), the CSI is computed as:

$$CSI = \begin{cases} \frac{(SoC_a - SoC_{ini})}{(SoC_d - SoC_{ini})}, & if\, SoC_a < SoC_d \\ 1, & else \end{cases} \tag{2}$$

4.4. Modeling Customer Behavior

In our basic scenario, we modeled EV users via their arrival time t_A and departure time t_D at the charging station, plus their average energy demand (specifically, the SoC at arrival SoC_{ini}). We also assumed a fixed target SoC of 80%. For each user, the specific values for t_A, t_D, and SoC_{ini} were drawn from a normal distribution with mean 9:00 for arrival time, 17:00 for departure time, and 30% for initial state of charge at arrival. The standard deviations were selected as follows: 60 min for t_A and t_D and 10% for SoC_{ini}.

Before the start of the simulation, we first generated a set of N_E users with the above given statistics, so that user (EV) u had an individual value $t_A(u)$, and so on. Based on these data, we calculated the charging demands for each EV and each day by determining arrival time, departure time, and initial SoC, again from a Gaussian distribution. The mean of the distribution was taken from the user characterization for each u, e.g., $t_A(u)$, with a variance that was half the inter-user variances (i.e., 30 min for t_A and t_D and and 5% for SoC_{ini}).

This means that there were both variations in the average values between users, as well as daily variations. Note that in addition to charging demands, weather conditions will differ, so that typical CSI values will vary over different days. The process for computing the daily charging demands is depicted in Figure 2. As one can see, it is also part of the user behavior that the vehicle can also be loaded externally to the company and that the user has to wait when there are not enough free charging stations. This also has an effect on t_A and the CSI calculation.

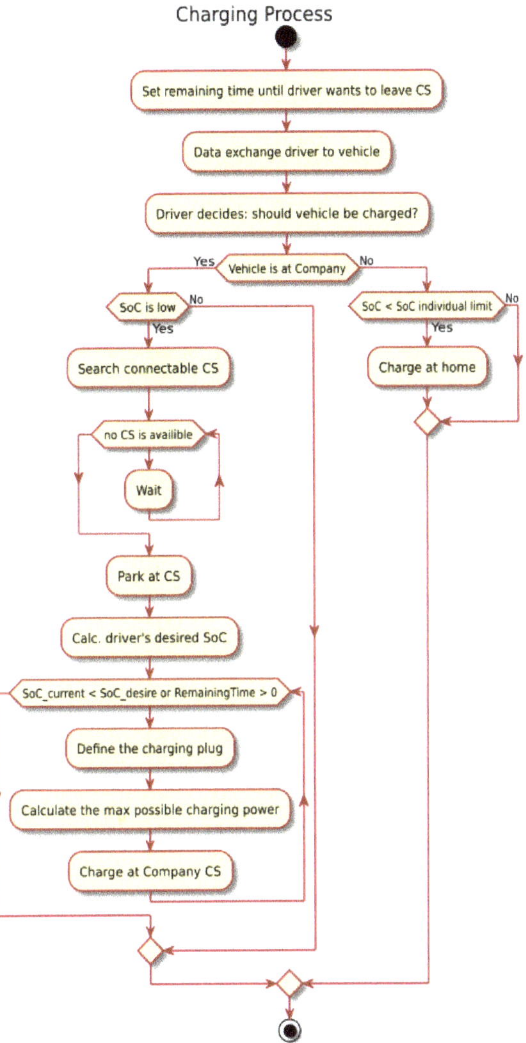

Figure 2. Modeling of the charging process in the MAS.

4.5. Extension to Weather Dependency

As a first extension of the basic scenario, we added a sensitivity in the charging demands to external conditions, which become visible at run-time. Specifically, we added a change in electrical energy demand of the vehicles dependent on the ambient temperature T. For simplicity, a simple relationship between outside temperature T and the additional vehicle energy consumption E_{Spec} due to air conditioning, heating, and battery behavior was assumed (see [23,24] for more information):

$$E_{Spec} = \begin{cases} -1.4458T + 21.255, & (-20 < T \leq 14) \\ 0, & (14 < T \leq 20) \\ 1.25T - 25, & (21 < T \leq 40) \end{cases} \tag{3}$$

This means that vehicles with a lower SoC arrive at the charging point and more energy is needed to fully charge the battery. The relation between energy demand and outside temperature is shown in Figure 3.

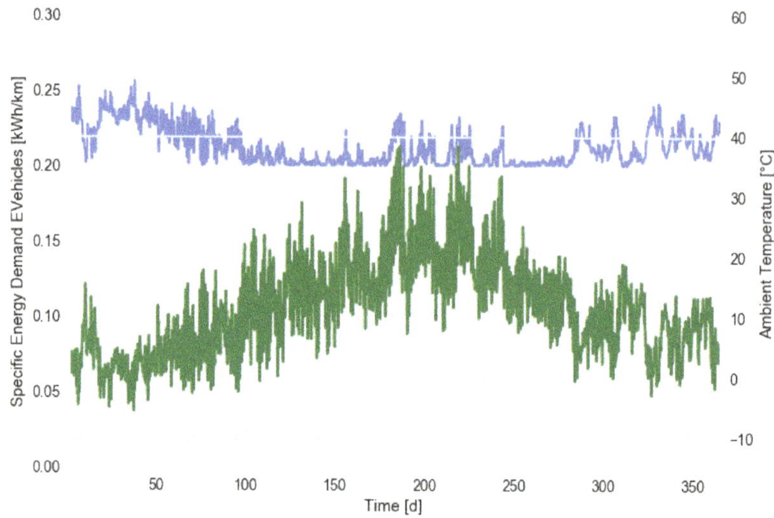

Figure 3. Temperature-dependent energy demand (in kWh/km) of EVs.

4.6. CSI Response Scenario

In the third scenario, we modeled a CSI sensitivity for all agents. After every charging process, the MAS computes the CSI value for the last charging session of each user. If for EV u, this values is under 80%, more than $n_f = 3$ times, the agent u will not use the charging system anymore due to frustration. Obviously, this is a very simple scenario, but visualizes the potential impact of adaptive customer behavior.

5. Results

We simulated the three scenarios defined above for a complete year using weather data for a city in Germany. Now, we want to take a look at the total charged energy in the CSI scenario for different values of P_{Base} in Figure 4. We see that PV power on its own was not sufficient since the transferred energy increased substantially when P_{Base} was increased. However, the selected value of $P_{Base} = 30$ kW covered 87% of the maximum charging energy amount.

Looking at the impact of different scenarios on the total charged energy in Figure 5, we observed a small increase in the temperature-dependent usage scenario and a strong decline to about 65% of the base demand for the CSI scenario.

The daily charging power for the three different scenarios is shown in Figure 6.

The dependency of the number of daily charging processes is shown in Figure 7 for different scenarios and in Figure 8 for different values of P_{Base}. We see that in the CSI scenario, the number of charging processes fell strongly in the first few months. In the final month (December), the mean number of charging processes for the CSI scenario was 47% lower than for the base scenario. An increasing grid energy supply (Figure 8) increased the number of long-term charging processes (customers), but even unlimited charging power did not avoid the effect of disappointed customers. This was due to charging requirements that were sometimes too demanding given the limited charging power of the CS or their EVs.

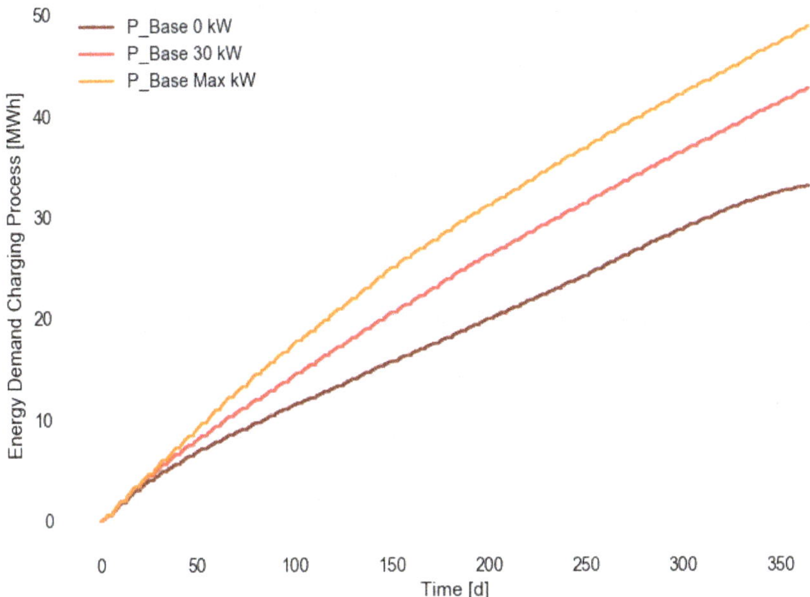

Figure 4. Accumulated charging energy over time for different values of P_{Base} for the CSI scenario.

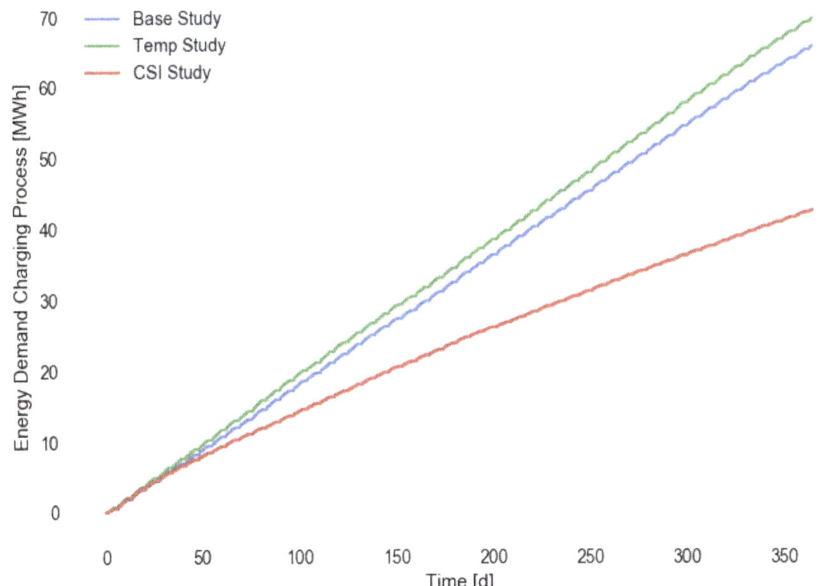

Figure 5. Accumulated energy demand over time for different agent scenarios (with $P_{Base} = 30$ kW).

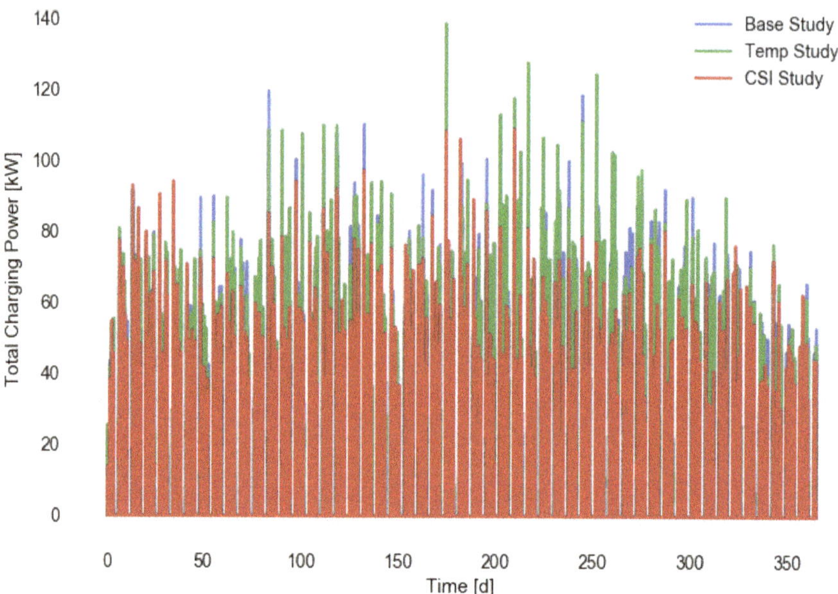

Figure 6. EV charging power (for all vehicles) for different scenarios.

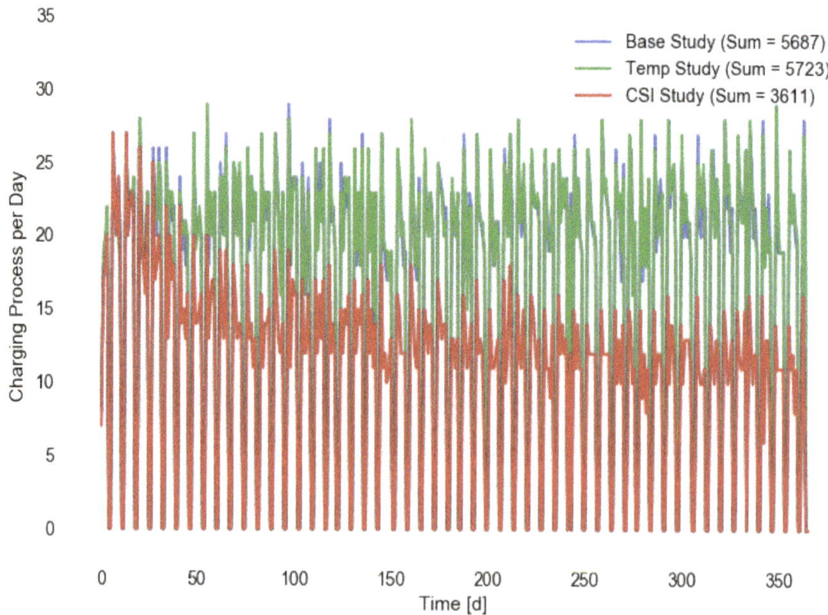

Figure 7. Number of daily charging processes for different scenarios.

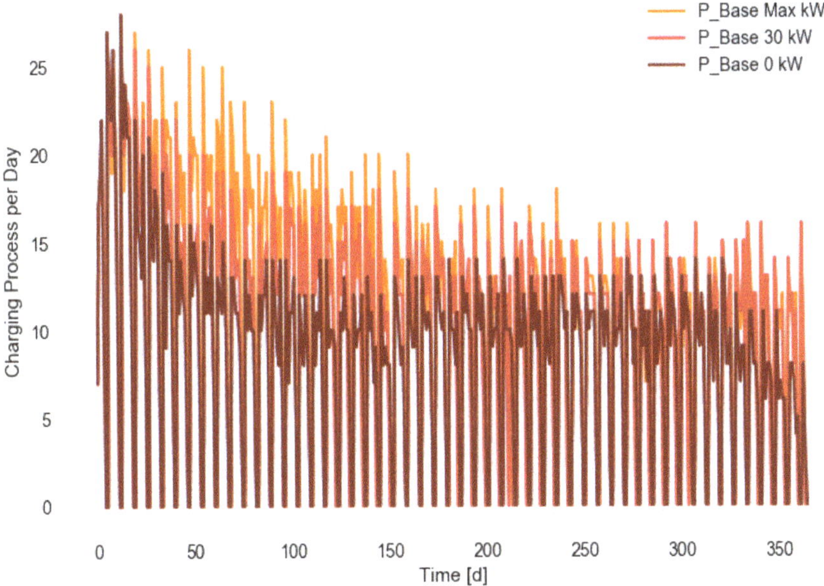

Figure 8. Number of charging processes each day in the CSI sensitive scenario for different values of P_{Base}.

A boxplot of CSI values for all three scenarios is shown in Figure 9. Note that this plot only shows charging processes with a CSI of <1.0. In fact, 93–98% (depending on scenario) of all charging processes provide full customer satisfaction. Differences for the three scenarios were very small, with the CSI scenario even having the best average values and the Temp scenario the worst values. The changes for the CSI scenario when adapting P_{Base} are shown in Figure 10. We note an increase in CSI values with more base charging power (taken from the grid). However, due to limited charging power, customer satisfaction might still be suboptimal even when a very large grid supply is available.

Finally, we study the total number of unsatisfying (CSI < 0.8) charging sessions. Figure 11 shows variations for different scenarios and Figure 12 the impact of P_{Base} in the CSI scenario. Assuming that a CSI level below a threshold (0.8 in our simulations) indicates that the corresponding charging session led to customer frustration, these plots show the development in the number of weak charging sessions. These numbers might for example in real usage be related to the number of customer complaints. It is interesting to observe that in the CSI scenario, the number of low CSI charging sessions increased much slower than in the other two scenarios (Figure 11). After about two months, there was only a moderate increase in the total number, while in the Base and Temp scenario, those numbers increased much faster. This was due users with more difficult charging requirements switching to a different charging location, leaving the remaining users with a more suitable solution. A possible implication of CSI-sensitive behavior compared to the other scenarios is that a charging operator might substantially overestimate the required effort for handling customer complaints.

As expected, the temperature-dependent charging demand also increased the number of low CSI sessions relative to the baseline. A quite surprising finding in Figure 12 was the low difference in the number of final low-quality charging sessions at the end of the year for different P_{Base} values (101 for 0 kW, 76 for 30 kW, and 68 for Max power (1 MW)). In terms of service efficiency (providing high CSI level charging sessions), even large variations in P_{Base} had a rather small impact. In contrast, the response of the customers seemed to add actually some self-regulating feature to the overall system.

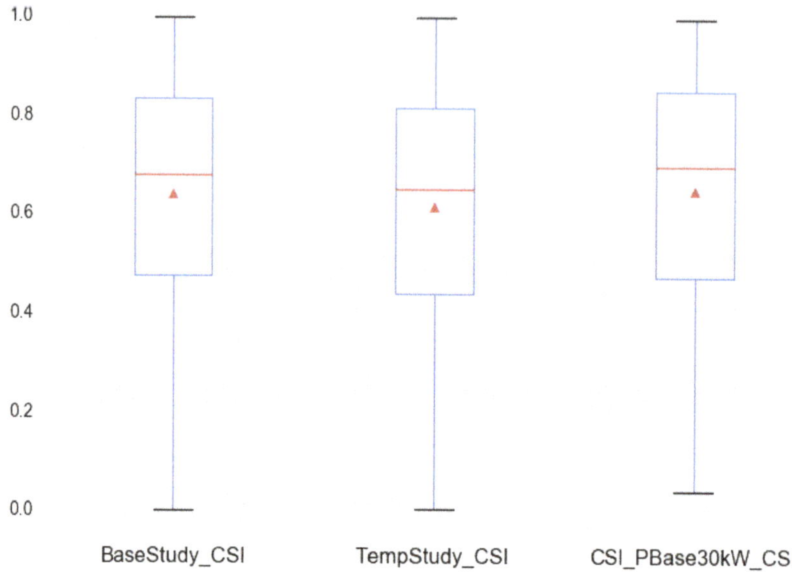

Figure 9. Average CSI values over a year for different scenarios for the standard (30 kW P_{Base}) setting. Note that only charging processes with a CSI value of less than 1.0 are reflected here.

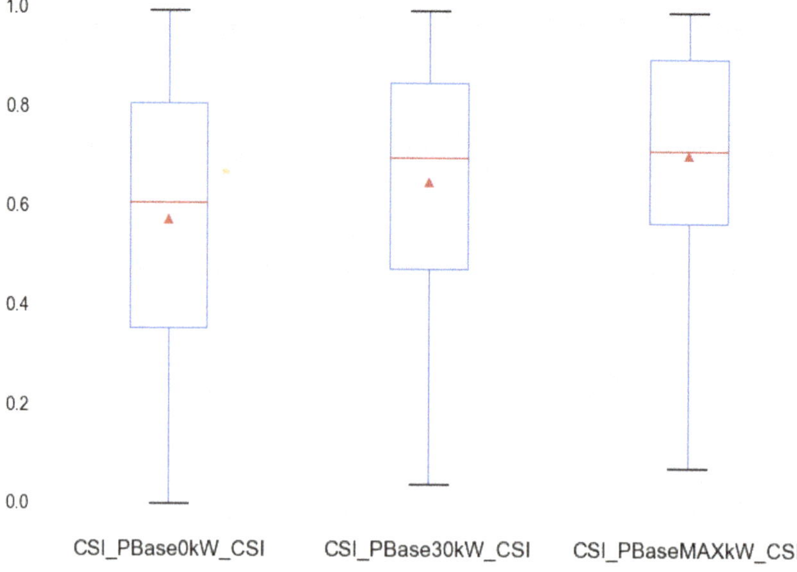

Figure 10. Boxplot of yearly CSI statistics for different values of P_{Base} in the CSI-sensitive scenario. Note that only charging processes with a CSI value of less than 1.0 are reflected here.

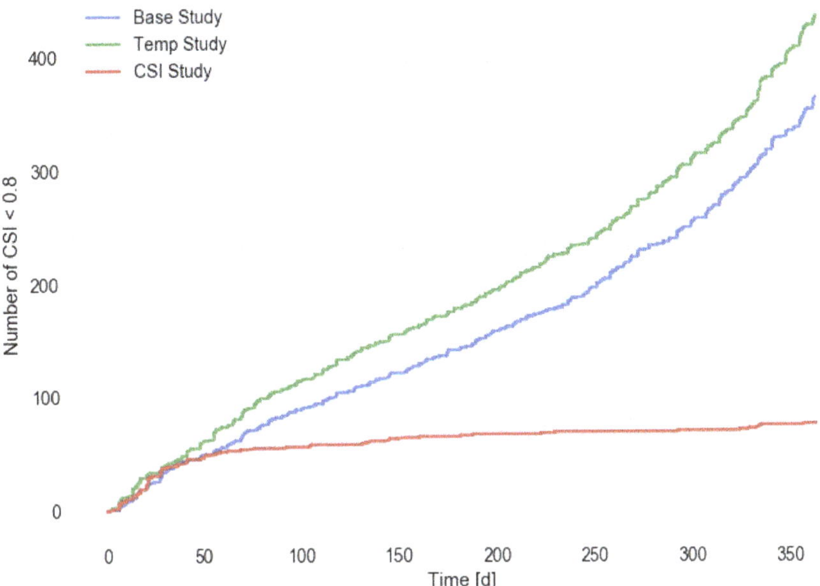

Figure 11. Accumulated number of low (<0.8) CSI charging processes for different scenarios (with $P_{Base} = 30$ kW).

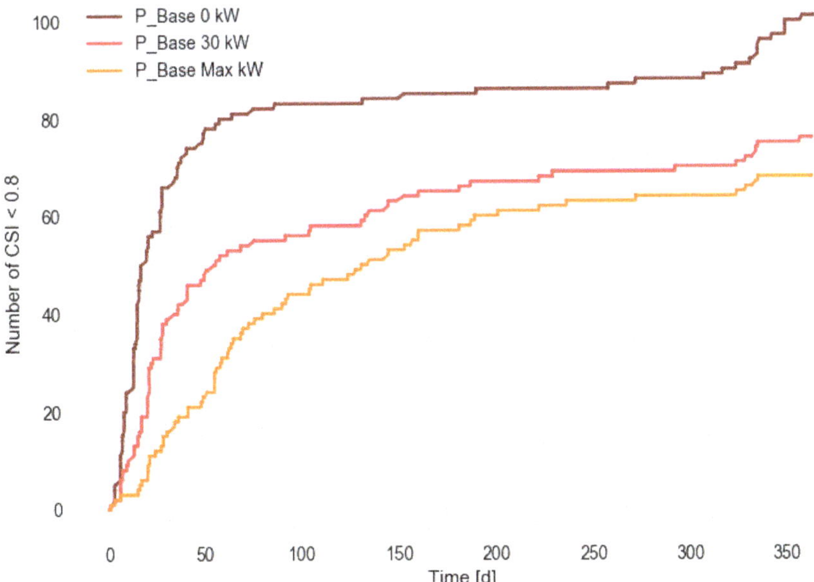

Figure 12. Accumulated number of low (<0.8) CSI charging processes in the CSI-sensitive scenario for different P_{Base} values.

6. Discussion

In this work, we used a multi-agent system approach for the assessment of the performance of a charging system. EV users were modeled at different levels of detail. In contrast to standard pre-calculated, table-based approaches, an MAS can allow agents (users) to respond to external factors like weather conditions, but also to the results of previous charging sessions.

In an example study, we investigated two different effects: an outside temperature-dependent charging energy demand and a customer satisfaction- sensitive charging behavior. The temperature-sensitive scenario was selected as an example of a technical dependency of the charging demands on external (environmental) conditions, while the CSI-sensitive scenario was considered as a very basic human response pattern that has been often ignored in the analysis of charging systems. The two scenarios led to substantial differences in the key performance values of the system (charged energy, number of daily charging sessions). The CSI-sensitive customer scenario even led to qualitatively different results, although low quality charging processes were very rare. It also turned out that increasing the base power from the grid had a positive influence on the charging performance, but did not generally avoid the issues with unsatisfied customers. It might therefore be better to accept a certain number of discontented customers to provide a better and less expensive service to the others.

In prior work [25,26], we proposed a dynamic pricing scheme, where charging station operators offer different prices to customers depending on the customers' flexibility. The framework proposed in this paper could be used to better understand the impact of different pricing schemes on user behavior. This is especially relevant if additional constraints like fairness need to be accurately modeled [27].

Another interesting venue is to analyze the resulting real-world multi-agent system consisting of grid operators, aggregators, EVs, customers, and other agents, from a more fundamental system level along the lines of [28].

We conclude that a more complex user modeling, as facilitated by MAS, is an essential component in any simulation-based development and testing environment for smart charging systems. A key issue that needs to be solved is how to obtain the necessary user data to feed the agent models in the MAS. This is a specifically difficult topic for the energy domain, where due to rapidly changing structures, customer behavior is likely to change over time. Furthermore, not much data on general users of EVs (in contrast to early adopters and members of research institutions) is available.

7. Materials and Methods

Due to technical, license, and IP-property-related issues, we are unable to provide the underlying simulation framework. However, we are happy to share simulation output data on demand.

Author Contributions: Study conceptualization by T.R. and T.E.; methodology developed by T.E. and T.S.; software developed by T.E. and R.U.; formal analysis by T.R., T.E., and R.U.; writing by T.R. and T.E.; supervision by R.U.

Funding: This work was funded by the Honda Research Institute Europe GmbH.

Acknowledgments: We thank Martin Stadie and Andre Burger from Honda R&D Europe for providing all the necessary information to build the simulation model.

Conflicts of Interest: The authors declare that there is no conflict of interest.

Abbreviations

The following abbreviations are used in this manuscript:

MAS   multi-agent system
EV   electric vehicle
SoC   (battery) state of charge
CS   charging station

CSI   customer satisfaction indicator
FMI   functional mockup interface
FMU   functional mockup unit

References

1. Esterle, K. Einfluss der Nutzung elektrischer Fahrzeuge auf ein regeneratives Gebäudeenergiesystem. Master's Thesis, TU Dresden, Dresden, Germany, 2014.
2. Mehta, R.; Srinivasan, D.; Trivedi, A. Optimal charging scheduling of plug-in electric vehicles for maximizing penetration within a workplace car park. In Proceedings of the 2016 IEEE Congress on Evolutionary Computation (CEC), Vancouver, BC, Canada, 24–29 July 2016; pp. 3646–3653. [CrossRef]
3. Daina, N.; Sivakumar, A.; Polak, J.W. Electric vehicle charging choices: Modelling and implications for smart charging services. *Transp. Res. Part C Emerg. Technol.* **2017**, *81*, 36–56. [CrossRef]
4. Canizes, B.; Soares, J.; Costa, A.; Pinto, T.; Lezama, F.; Novais, P.; Vale, Z. Electric Vehicles User Charging Behaviour Simulator for a Smart City. *Energies* **2019**, *12*, 1470. [CrossRef]
5. Franke, T.; Krems, J.F. Understanding charging behaviour of electric vehicle users. *Transp. Res. Part F Traffic Psychol. Behav.* **2013**, *21*, 75–89. [CrossRef]
6. Xydas, E.; Marmaras, C.; Cipcigan, L.M. A multi-agent based scheduling algorithm for adaptive electric vehicles charging. *Appl. Energy* **2016**, *177*, 354–365. [CrossRef]
7. Rolim, C.C.; Gonçalves, G.N.; Farias, T.L.; Rodrigues, Ó. Impacts of electric vehicle adoption on driver behavior and environmental performance. *Procedia Soc. Behav. Sci.* **2012**, *54*, 706–715. [CrossRef]
8. Maia, R.; Silva, M.; Araujo, R.; Nunes, U. Electric vehicle simulator for energy consumption studies in electric mobility systems. In Proceedings of the 2011 IEEE Forum on Integrated and Sustainable Transportation Systems, Vienna, Austria, 29 June–1 July 2011; pp. 227–232.
9. Galus, M.D.; Dobler, C.; Waraich, R.A.; Andersson, G. *Predictive, Distributed, Hierarchical Charging Control of PHEVs in the Distribution System of a Large Urban Area Incorporating a Multi Agent Transportation Simulation*; Technical Report; ETH: Zürich, Switzerland, 2011.
10. Sweda, T.; Klabjan, D. An Agent-Based Decision Support System for Electric Vehicle Charging Infrastructure Deployment. In Proceedings of the 2011 IEEE Vehicle Power and Propulsion Conference, Chicago, IL, USA, 6–9 September 2011.
11. Hu, J.; Saleem, A.; You, S.; Nordström, L.; Lind, M.; Ostergaard, J. A multi-agent system for distribution grid congestion management with electric vehicles. *Eng. Appl. Artif. Intell.* **2015**, *38*, 45–58. [CrossRef]
12. Laclavik, M.; Dlugolinsky, S.; Seleng, M.; Kvassay, M.; Schneider, B.; Bracker, H.; Wrzeszcz, M.; Kitowski, J.; Hluchy, L. Agent-Based Simulation Platform Evaluation in the Context of Human Behavior Modeling. In Proceedings of the 10th International Conference on Advanced Agent Technology, Taipei, Taiwan, 2–6 May 2011; Springer: Berlin/Heidelberg, Germany, 2012; pp. 396–410.
13. Athavale, S.; Balaraman, V. Human Behavioral Modeling for Enhanced Software Project Management. In Proceedings of the CONSEG 2013 7th CSI International Conference on Software Engineering, San Francisco, CA, USA, 18–26 May 2013.
14. Railsback, S.F.; Lytinen, S.L.; Jackson, S.K. Agent-based Simulation Platforms: Review and Development Recommendations. *Simulation* **2006**, *82*, 609–623. [CrossRef]
15. ESI-ITI. SimulationX 4.0. Available online: http://www.simulationx.com/ (accessed on 22 July 2019).
16. Fritzson, P.; Bunus, P. Modelica—A General Object-Oriented Language for Continuous and Discrete-Event System Modeling. In Proceedings of the 35th Annual Simulation Symposium, San Diego, CA, USA, 14–18 April 2002; pp. 365–380.
17. Blochwitz, T.; Otter, M.; Åkesson, J.; Arnold, M.; Clauss, C.; Elmqvist, H.; Friedrich, M.; Junghanns, A.; Mauss, J.; Neumerkel, D.; et al. The Functional Mockup Interface for Tool independent Exchange of Simulation Models. In Proceedings of the 8th International Modelica Conference, Dresden, Germany, 20–22 March 2011.
18. Rodemann, T. A Comparison of Different Many-Objective Optimization Algorithms for Energy System Optimization. In *Applications of Evolutionary Computation*; Kaufmann, P., Castillo, P., Eds.; LNCS 11454; Springer: Berlin/Heidelberg, Germany, 2019; pp. 1–16.

19. Rodemann, T.; Unger, R. Smart Company Digital Twin-Supporting Controller Development and Testing Using FMI. In *Proceedings of the Japanese Society of Automotive Engineers Spring Meeting*; JSAE: Yokohama, Japan, 2018.
20. Weiß, G. Agentenorientiertes Software Engineering. *Informatik Spektrum* **2001**, *24*, 98–101. [CrossRef]
21. Weiss, G. *Multiagent Systems*; The MIT Press: Cambridge, MA, USA, 2013.
22. Rodemann, T.; Kitamura, K. Simulation-based Design and Evaluation of a Smart Energy Manager. In *Lecture Notes in Computer Science—Computer Aided Systems Theory*; Springer: Berlin/Heidelberg, Germany, 2019.
23. Hülsebusch, D.; Schwunk, S.; Caron, S.; Propfe, B. Modeling and simulation of electric vehicles—The effect of different Li-ion battery technologies. In Proceedings of the 25th World Battery, Hybrid and Fuel Cell Electric Vehicle Symposium & Exhibition (EVS 25), Shenzhen, China, 7–9 November 2010.
24. Schücking, M.; Jochem, P.; Fichtner, W.; Wollersheim, O.; Stella, K. Influencing factors on specific energy consumption of EV in extensive operations. In Proceedings of the 29th International Electric Vehicle Symposium 2016 (EVS29), Montréal, QC, Canada, 19–22 June 2016; Volume 1, pp. 195–203.
25. Limmer, S.; Rodemann, T. Multi-objective Optimization of Plug-in Electric Vehicle Charging Prices. In Proceedings of the IEEE SSCI 2017, Honolulu, HI, USA, 27 November–1 December 2017; pp. 2853–2860.
26. Limmer, S.; Rodemann, T. Peak Load Reduction through Dynamic Pricing for Electric Vehicle Charging. *Int. J. Electr. Power Energy Syst.* **2019**, *113*, 117–128. [CrossRef]
27. Limmer, S.; Dietrich, M. Optimization of Dynamic Prices for Electric Vehicle Charging Considering Fairness. In Proceedings of the 2018 IEEE Symposium Series on Computational Intelligence (SSCI), Bangalore, India, 18–21 November 2018; pp. 2304–2311.
28. Sadik, A.R.; Urban, B. Combining Adaptive Holonic Control and ISA-95 Architectures to Self-Organize the Interaction in a Worker-Industrial Robot Cooperative Workcell. *Future Internet* **2017**, *9*, 35. [CrossRef]

Article

Wireless Sensor Network Energy Model and Its Use in the Optimization of Routing Protocols

Carolina Del-Valle-Soto [1,*], Carlos Mex-Perera [2], Juan Arturo Nolazco-Flores [3], Ramiro Velázquez [4] and Alberto Rossa-Sierra [1]

[1] Facultad de Ingeniería, Universidad Panamericana, Álvaro del Portillo 49, Zapopan, Jalisco 45010, Mexico; lurosa@up.edu.mx

[2] ITAM, Rio Hondo 1, Ciudad de México 01080, Mexico; carlos.mex@itam.mx

[3] Tecnologico de Monterrey, Campus Puebla, Vía Atlixcáyotl No. 5718, Reserva Territorial Atlixcáyotl, Puebla 72453, Mexico; jnolazco@tec.mx

[4] Facultad de Ingeniería, Universidad Panamericana, Josemaría Escrivá de Balaguer 101, Aguascalientes 20290, Mexico; rvelazquez@up.edu.mx

* Correspondence: cvalle@up.edu.mx; Tel.: +52-33-13682200 (ext. 4245)

Received: 20 December 2019; Accepted: 30 January 2020; Published: 7 February 2020

Abstract: In this study, a Wireless Sensor Network (WSN) energy model is proposed by defining the energy consumption at each node. Such a model calculates the energy at each node by estimating the energy of the main functions developed at sensing and transmitting data when running the routing protocol. These functions are related to wireless communications and measured and compared to the most relevant impact on an energy standpoint and performance metrics. The energy model is validated using a Texas Instruments CC2530 system-on-chip (SoC), as a proof-of-concept. The proposed energy model is then used to calculate the energy consumption of a Multi-Parent Hierarchical (MPH) routing protocol and five widely known network sensors routing protocols: Ad-hoc On-demand Distance Vector (AODV), Dynamic Source Routing (DSR), ZigBee Tree Routing (ZTR), Low Energy Adaptive Clustering Hierarchy (LEACH), and Power Efficient Gathering in Sensor Information Systems (PEGASIS). Experimental test-bed simulations were performed on a random layout topology with two collector nodes. Each node was running under different wireless technologies: Zigbee, Bluetooth Low Energy, and LoRa by WiFi. The objective of this work is to analyze the performance of the proposed energy model in routing protocols of diverse nature: reactive, proactive, hybrid and energy-aware. Experimental results show that the MPH routing protocol consumes 16%, 13%, and 5% less energy when compared to AODV, DSR, and ZTR, respectively; and it presents only 2% and 3% of greater energy consumption with respect to the energy-aware PEGASIS and LEACH protocols, respectively. The proposed model achieves a 97% accuracy compared to the actual performance of a network. Tests are performed to analyze the consumption of the main tasks of a node in a network.

Keywords: energy consumption; routing protocols; performance parameters; Wireless Sensor Network (WSN)

---

1. Introduction

Presently, houses, buildings, parks and, cities in general, involve several electronic devices working with different wireless technologies. The type of application determines the appropriate combination of standards and protocols to be used. One of the main characteristics of these wireless devices is their power requirements.

Wireless Sensor Networks (WSNs) involve devices characterized by small nodes, low energy consumption, limited battery life, low task processing, and low storage capacity. These self-configuring networks are easy to implement and to deploy. In these networks, communications through channels with multiple interferences and computing capabilities to operate at low energy are assessed. Sensor networks should exhibit an optimal performance with reduced delays and provide reliable information with a minimum energy consumption in order to provide valuable information for long periods [1]. However, energy consumption may become a major issue because of the low-battery power. The life span of the nodes should be as long as possible to avoid constant human intervention due to the harsh environment of some of their applications, such as in the study of natural behavior, risk areas, medical industry, domotics, agriculture, battlefields, and home networks [2].

Considerations on energy consumption are critical in sensor networks because their implementation must be simple, enduring, and resilient to topology or configuration changes. All these factors significantly influence the energy expenditure of a network and are represented by its performance parameters.

In this study, an efficient energy model for sensor networks is implemented with the aim of quantifying energy consumption during the execution of the main tasks of a node within a network [3]. The types of energy considered by this model correspond to the following phases of operation: turning on, channel auditing, receiving packets, sending packets, switching activities, microcontroller processing, and turning off. The novelty of this model is that it features a simple design scheme capable of characterizing energetic behavior against possible network anomalies in which consumption levels exceptionally increase, as it can be seen in [4]. In the proposed model, energy is considered to be an indicator of a typical behavior for a network exposed to interference attacks. In addition, it is an easy scheme to implement in a node, which provides reliable responses to modifications in node behavior. In this way, it is also possible to optimize energy consumption for specific node activities as well as for the overall network performance. The model allows for scalability and demonstrates the main reaction modes of a network node.

This paper intends to test the proposed energy model and observe its repercussions on proactive and reactive sensor network protocols. The analysis is described quantitatively by observing the performance metrics that would positively or negatively affect such model. This is where the contribution of this work becomes relevant: the proposed model quickly shows changes in the network performance, its implementation is simple, and it does not represent higher processing consumption. The proposed model is then compared against the performance of network sensors under some widely known protocols: Ad hoc On demand Distance Vector (AODV) [5], Dynamic Source Routing (DSR) [6], ZigBee Tree Routing (ZTR) [7], Low Energy Adaptive Clustering Hierarchy (LEACH) [8] and, Power Efficient Gathering in Sensor Information Systems (PEGASIS) [9]. These protocols will also be compared against the Multi-Parent Hierarchical (MPH) routing protocol proposed, designed, and implemented by the authors in a previous work [10].

There are also other protocols analyzed in the literature, such as Cluster-based Energy Efficient Location Routing Protocol (CELRP), which is a hierarchical protocol with nodes distributed in clusters and arranged in quadrants. Each quadrant contains two clustering, which would be like the master nodes, and other nodes transmit data with two hops data transmission. Another similar protocol is Position Responsive Routing Protocol (PRRP), which is more energy efficient. This protocol makes a choice of the cluster head based on distance from the sink, energy level, and the average distance of neighboring nodes from the candidate master node. PRRP is similar to the LEACH protocol in which any node can communicate with the sump and the data transmission mechanism is the time-based schedule. In PRRP, the number of nodes of the branches of the hierarchical tree and the distance from the non-leaf node is smaller compared to LEACH and CELRP. This makes energy conservation candidate for optimization [11]. The PRRP protocol dramatically increases data transfer and provides a better solution to the routing problem focused on energy efficiency, due to the efficient selection and distribution of gateways. Another important protocol to mention is the Energy-Efficient data

Routing Protocol (EERP) for WSN [12], which selects a set of good roads, and chooses the one based on the node state and the road cost function. In EERP, each node has several neighbors through which packets can be routed to the base station. A node bases its routing decision on two metrics: status and cost function.

In this study, AODV, DSR, ZTR, LEACH, and PEGASIS are quantitatively compared and assessed based on several efficiency metrics that analyze how these routing protocols optimize energy through various schemes in order to find the best routes in the shortest possible time. As the hierarchy algorithms, such as the ZTR, denote simple and fast routing that reduce network overloads, they are reliable and have a distributed addressing scheme that only permits neighbor tables, not long, and elaborated routing tables. The performance of WSN is closely related to that of the routing protocol, because routes can vary dynamically over time. Energy-aware protocols such as LEACH and PEGASIS seek to increase the lifetime of the network. They propose to find sub-optimal paths to allow a more equitable distribution of the network's energy consumption. Hierarchical protocols such as ZTR and MPH have advantages in terms of scalability and efficiency in communications. Particularly for WSN, nodes with higher energy can be used to process and send information, while those with lower energy are used to monitor the environment and send the information to the node with greater energy capacity. Finally, proactive type protocols, which establish routes before there is a real traffic demand, are suitable for real-time traffic, since they have low latency; however, they waste bandwidth due to periodic updates and they are not energy efficient. The MPH protocol is a hybrid protocol, i.e., it is a combination of reactive and proactive nature protocols. The AODV and DSR protocols are two protocols widely recognized in WSN for their rationality and the ZTR protocol is a proactive protocol par excellence. Therefore, we believe that the comparison between protocols of natures of the same type is relevant. In addition, we wanted to complement our study with energy-aware protocols because the main objective of this work is to demonstrate and analyze the higher energy costs in the sensors, according to the type of tasks they perform on the network.

This paper proposes a simple energy model, which quickly shows changes in the network performance, its implementation is simple, and it does not represent higher processing consumption. In the WSN literature, there are few energy models [13,14] and some energy-aware protocols that seek to optimize the energy of networks. The need for an energy model that impacts the performance metrics of a network is an advantage that not all models exhibit. This indicates that we can know, according to each type of task that the node performs, what is the major and minor impact on parameters such as: resilience, overhead, packet retransmissions, listening retries to the communication channel, delay, and many others.

The energy consumption problem is not the same for all network nodes [15]. This is due to the fact that there are several collector nodes that cluster information around them. This coordinating collector node is robust, with its own energy supply and with greater processing resources than other nodes in the network. Consequently, this node has the capacity to process all the information gathered from the nodes of the network and subsequently, obtains results when assessing the information received. When there are one or more collector nodes and there are nodes nearby that forward all network traffic, they are more likely to exhaust faster their energy. This problem is known as the energy hole problem [16] and generates a high amount of packet losses, which will be represented by collisions. Unequal energy depletion causes the expiration time to unexpectedly generate loss of information from the network.

A possible solution to this problem, as per the literature, is the creation of groups to promote network scalability and the problem of zoning [17]. The analysis of the network areas is distributed in concentric rings to stimulate traffic between nodes as they approach their destination. The authors in [18] study variables such as constant bit rate, where the nodes are uniformly and randomly distributed. Performance parameters are analyzed to establish relationships between the different rings located near and far away from the collector node. In this study, the concept of node zones is used to observe energy repercussions based on accurate and reliable performance metrics that directly influence energy expenditure and are presented as reliable evidence of network changes or anomalies. The link load imbalance problem is addressed through tree-type topology protocols aimed at minimizing packet delays and the number of jumps in the path to the root node or collector node. This imbalance is reflected as a non-uniform energy expenditure on the nodes. This failure is compensated in reactive protocols by adding more links, but this is prone to greater delays or outdated routes and their maintenance may become a challenge [19]. The work in [20] shows a method for balancing network traffic and ensuring uniform use of destination node routes. Nezhad et al. [21] propose a protocol in which the collector node has a global view of the network topology to maximize the life span of each node and the use of a load balancing algorithm to select the best routes.

Herein, some energy consumption performance metrics are proposed and used to compare the MPH routing protocol against other known sensor network protocols: AODV, DSR, ZTR, LEACH, and PEGASIS, all as per the IEEE standard 802.15.4/ZigBee. Node shutdown tests are performed to study how the network behaves according to the routing protocol implemented and its response capabilities [22]. The results of the simulation show how performance, reliability, and energy consumption are affected within the communications network. In this light, MPH is shown to be an efficient protocol as it presents the best performance among the protocols under evaluation. Compared to AODV, MPH routing exhibits a 26.9% decrease in overall energy consumption and a 41.2% increase in the protocol's ability to recover the topology after a failure event. In addition, an energy model for the CC2530 chip is proposed and used in the simulations of the four protocols aforementioned, resulting in a 16%, 13%, and 5% reduction in energy consumption for the MPH routing when compared to AODV, DSR, and ZTR, respectively. The proposed model allows us to determine that between MPH, LEACH and PEGASIS there is only a difference of 3% and 2% energy savings for the last two protocols. Thus, the model analyzes the energy impact of each type of energy for optimization of the algorithm in various protocols of the literature. These protocols and the energy model are implemented in an event simulator programmed in C++. The proposed energy model is implemented in the simulator for each routing protocol to observe the impact of the performance parameters and how they influence energy consumption in each protocol. This proposal resembles the MPH routing protocol [10], which is a hybrid that combines proactive and reactive protocols and energy conservation properties at the same time. It is used in this work as a protocol with hierarchical topology for optimizing the sending of information, maintaining a smaller number of routes, and monitoring them to determine whether they remain valid or have become obsolete [23].

Figure 1 shows a conceptual scheme for a hierarchical topology with a sink node or information collector. This scheme is widely used in WSN for low power consumption [24]. The problem described here is that energy is not distributed equally throughout the topology and there are different energy levels called crowns [25]. Energy increases in the nodes that form information bottlenecks. Therefore, in this work we classify the types of energy to establish compensations in the busiest nodes of the network. This scheme overviews the energy consumption of a many-to-one network. The coordinator or sink node is the sole destination of the network nodes (there may be more than one coordinator node), which makes it a single failure point. The nodes send traffic to the coordinator node. As the nodes are closer to the coordinator, they have to forward traffic from other network nodes. Therefore, with this scheme, we want to represent an ideal scenario of energy behavior of the nodes from the coordinator node to the nodes farthest from it. Being the darkest color the higher energy consumption nodes and the lightest, the lower energy consumption ones.

1.1. Motivation

This work presents a simple energy model based on the detailed analysis of the main types of energy consumed during the different tasks performed by sensor within a network while in active mode. This model is capable of performing a detailed breakdown of such types of energy according to the node hierarchy in the network considering at the same time both high and low consumption sources. This energy model is characterized by being simple and providing evidence of the energy used by the main tasks of a node in the network. In addition, it reflects the impact of each type of energy on the performance metrics of a network, with which we could optimize routing protocols according to the conditions of the network and its local expenditure (at node level).

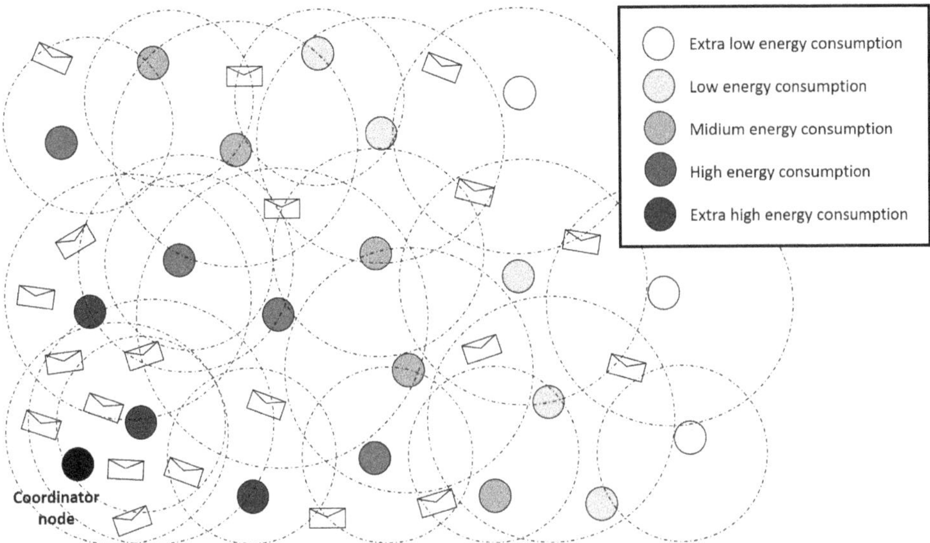

Figure 1. Conceptual energy consumption model in a hierarchical topology.

The results in this proposal are established by assigning a node hierarchy based on the traffic load due to the node's proximity to the coordinator node. The novelty in this proposal is the possibility of obtaining the skeleton of each energy expenditure in the network and establishing optimizations for higher consumptions. Furthermore, these network consumptions are classified by the proposed energy model according to their dependency in data packet sending. For example, if the unit's microcontroller is detected to be of high consumption, it can be replaced by a smaller one or for a chip with a better wireless technology. Similarly, if the consumption during data reception or transmission is high, it can be an indicator that the channel exhibits high interference and intermittent links. This way, this simple and easy to implement model intends to be the starting point for specific task node optimization methods and sleeping algorithms.

1.2. Scope of Routing Protocols in Sensor Networks

Most of the nodes in a sensor network have a limited power supply and do not have the ability to generate their own energy. Therefore, the design of energy efficient protocols is critical for the longevity of the network. A protocol for sensor networks must be configured in such a way that its operation does not require human attention. The IEEE 802.15.4 standard is used by virtually all wireless sensor devices today. Since a direct link between any node in the network and the coordinating node cannot necessarily be established, a multi-hop network topology and an algorithm are required to determine the route the messages follow. These are dynamic topologies with nodes that can stop operating due to physical failures or lack of batteries, with bandwidth restrictions, links with variable capacities, and equipment that can operate with energy restrictions. All these factors result in reconfigurations or unpredictable changes in the topology that handles the routing protocol. On the basis that many routes can communicate to a node with the base station or coordinating node, the objective of an energy-aware algorithm is the selection of those routes that maximize the lifetime of the network. In consequence, those routes composed of nodes that have greater autonomy are marked as preferred. Presently, some of the hierarchical protocols widely used in sensor networks are: LEACH (Low Energy

Adaptive Clustering Hierarchy) [26,27], PEGASIS (Power-Efficient Gathering in Sensor Information Systems) [28], TEEN (Threshold-sensitive Energy Efficient Protocols) [29], and PAMAS (Power-Aware Multi-access protocol with signaling) [30], among others.

LEACH is a cluster-based protocol that randomly selects a few nodes and treats them as master nodes. Its objective is to distribute the energy load evenly throughout the network. PEGASIS is an improvement of the LEACH protocol. Its main idea is to extend the life of the network by restricting node communication only to the closest neighbors and taking turns for communicating with the coordinating node. PEGASIS assumes that each node must be able to communicate with the base station directly. The TEEN protocol is used in applications where a rapid response is required against sudden changes in the parameters to be measured in the network. In TEEN, the nodes monitor the medium permanently, but the data transmission is sporadic. Since the transmission of messages consumes more energy than their capture, the consumption is lower than in proactive networks. PAMAS is a containment-based protocol where one of the main objectives is related to efficiency in energy consumption. This protocol adds separate channels for the control of RTS/CTS packets and data packets. "A busy tone" is attached to the signaling channel so that the nodes that do not participate in the communication turn their radio transceivers off. This way, PAMAS tries to avoid over-listening between neighboring nodes and does so by adding a second low power radio in their devices.

In communication networks, depending on the way the source creates a route to the destination, the routing protocols are classified into three groups: proactive routing protocols, reactive routing protocols, and hybrid protocols.

When nodes are under a reactive protocol, they ask for a single route as required. This implies a high latency for the first packet and some independence between the routes. Two common examples are the AODV and DSR protocols. These routing protocols are characterized by their reactive nature. That is, they both begin their search activity on demand routes. The difference between them is that DSR uses source routing, while AODV uses hop-by-hop routing by maintaining tables and sequence numbers in the destination nodes.

The AODV routing protocol is based on the routing efficiency of ad hoc wireless networks with a large number of nodes and uses a mechanism for route discovery in broadcast mode. AODV can transmit in unicast or multicast mode, uses bandwidth efficiently, and quickly responds to network changes avoiding network loops [5]. Each node in the network has a sequence number and a unique identifier in the network. This guarantees the absence of loops and avoids counting packets to infinity. To maintain and track routes to the neighbors, nodes periodically send *HELLO* messages. The nodes have a lifetime each time a node receives a packet from a neighbor. At this time, the neighbor's entry is updated in the routing table. If no entry has been defined for this neighbor, the node creates a new entry in the routing table. Therefore, information from the *HELLO* packet is used by neighbors to notify other nodes that the node itself is still active. This information is used by the neighbors to update the timers associated with that node or, alternatively, to disable the entries that are not responding. In fact, AODV maintains time-based states in each node's routing tables. A routing table entry expires if it has not been used recently. The timer function prevents using links which status has been marked as unknown for a long time. Some advantages of AODV are high reliability and low bandwidth costs. However, there are some disadvantages such as high complexity, additional calculations required, extra memory costs, and the fact that this protocol was designed to work in a network where there are no malicious nodes. In sum, it is not a secure protocol.

The DSR protocol is also a reactive protocol. This protocol routes from the source node including a header in the packets. This header indicates which nodes may need to be crossed to arrive at a destination since the originating node is responsible for calculating the complete route to the destination node. This process is called **Source Routing**. DSR does not require any periodic message. In this way, it reduces message overload. For this, when the origin node moves or the topology of the network changes, the algorithm perceives the modifications and adapts accordingly. In addition, DSR handles unidirectional links and asymmetric routes. Each node in the network has a cache memory that stores all the routes obtained through the discovery processes from the node itself, which may consume slightly additional processing. If there is no current route to a specific destination, the node starts a reactive route discovery just like AODV.

The route table or route cache is constantly monitored to detect invalid routes and repairs them as the network topology changes. This process is called **Route Maintenance**. DSR features some advantages such as that nodes can obtain multiple routes to a specific destination but requesting only one route. DSR allows the network to be completely self-configurable, without a specific architecture or topology. In addition, it is a good choice in scenarios where the number of mobility nodes is reduced. This protocol adapts quickly to routing changes when a node is frequently moving and decreases network overload.

Next, we study a hybrid (proactive and reactive) routing protocol previously proposed, called MPH, which creates a hierarchical network logical topology where node hierarchy is proactively given by its location in the tree. Basically, the root hierarchy (sink or collector node) has the highest hierarchy. When a node has a hierarchy level, it can only have links from parents or children, directly connected. The hierarchical topology minimizes the number of hops and optimizes the routes to the coordinating node. The advantage of this protocol is that it combines features of both proactive and reactive nature and presents redundancy without losing simplicity in the algorithm. The coordinator can also send packets to the network nodes. An origin routing approach is adopted so that traffic is sent from the collector to any node, because the coordinator node has more resources and capacity than the rest. This means that information may be easily collected from the network, such as sensor-generated data, link quality metrics, neighbor node tables, and other variables that may be used to locate routes and for performance analysis and network optimization. If a node sends a packet to the destination or collector node, it searches for its parents in its Neighbor Table, selects a route, and sends the message. This process is repeated hop by hop until the destination. The sink node knows the complete topology of the network through frequent query packets at times previously established in the network nodes. These packets are sent through the hierarchical branches and request each node to send its information to the coordinator node. This protocol takes advantage of the proactive controlled route maintenance but combines the agility of having more than one route per node. This makes it more versatile and adaptable to other topologies.

ZTR is a simple protocol that establishes parent-child links with the nodes always carrying information to their parent. This protocol features a tree topology, is easy to implement, fast, and is proactive. ZigBee networks require at least one full-function device. That is, a robust device acting as a network coordinator, but the final nodes of the star may have low specifications to reduce costs. Before transmitting to a channel, the node must measure the energy level in a specific channel. This measurement only indicates if the channel is busy, but the node is not able to identify whether that energy corresponds to another device under the IEEE 802.15.4 standard. Here, the child node would be the one that most recently entered the network. The parent node is the node that has given the child access to the network. In this way, parent-child links are created, but each child can only have one parent. Some of the advantages of ZTR are balance between cost per unit, battery expenditure, and implementation complexity to achieve an appropriate cost-performance ratio for the application.

1.3. Significance for Study of Energy in Wireless Sensor Networks

To evaluate the network performance, one may consider parameters that evidence proper network operation directly influencing the energy consumption of each node. There are local and global parameters. Global parameters display the total energy costs for the network considering each type of energy for each specific activity. In contrast, local parameters provide total energy consumption rates for a single node. This energy depends on the location of the node within the topology regardless of how near or far they are located from the coordinator node and how much traffic is transmitted through it [31].

An energy-efficient routing protocol decreases the consumption of the nodes by routing data through paths that display the least amount of energy. There are some special mechanisms to achieve this goal such as optimization of jumps to the destination node, maintenance of optimal and valid routes, reduction of transmission delays, and reduction of packet retransmissions and attempts to listen to the channel [32,33]. In this paper, we assess the aforementioned aspects to compare against four sensor network protocols.

Concerning the communication channel, it is a factor that significantly influences the energy consumption because the protocol executes a series of listening attempts to determine whether the channel is already busy with other information packets. The carrier senses multiple accesses with the collision avoidance (CSMA/CA) protocol [34] and works as follows: first, a node begins listening to the wireless channel and if it is free, the node begins transmitting. If the wireless channel is not free, the node recalculates a random delay, waits, and listens again. We are using the MAC-level protocol used for all extensions of 802.15.4 (including the original version), which is the CSMA/CA that guarantees a high data rate. A network recognition is being carried out at all times to check the status of the channel (carrier detection). Only when free, data can be sent. In the 802.11 standard, the physical layer polls the energy level over the radio frequency to determine whether or not there is transmission. If the channel is busy, a random timer starts (with a maximum of five back off periods), the timer only discovers time with free channel, transmits when it expires, and finally, if it does not receive ACK, it increases the back off. This metric is known as CSMA/CA retries. If these CSMA/CA retries are frequent, the channel is busy most of the time. Consequently, there might be several collisions due to overload. In addition, when the wireless channel is permanently busy with information packets, there are many collisions and retransmissions of packets. This fact influences energy consumption because the nodes spend more time and capacity retransmitting over and over.

In a network layer, overloads are an important factor that influence energy consumption. The efficiency of the routing protocol may also be measured by the number of packets the protocol needs to route to its destination. A protocol with many control packets will contribute to packet collisions and overall performance reduction. In terms of route discovery, in all the protocols considered, the nodes exhibit capacity to know their neighbors. In AODV and DSR, the nodes update the routes as required. However, in ZTR and MPH routing, nodes periodically update routes. Known neighbors helps nodes to establish valid destination routes to forward messages and to reduce the number of retransmission packets required. The metric named 'valid routes' consists of paths that are not damaged and these routes are immediately available and ready to be used for sending information. If nodes are able to handle several valid routes, it is more likely that packets will be forwarded and will not be retransmitted continuously, thus saving energy resources.

Another important metric is found in the tables where the routes to different network destinations are located. These routes may either be valid or obsolete. Hence, their maintenance is essential to prevent loss of information within the network. Network energy consumption is directly related to the complexity in the administration of routing or neighbor tables. As sensors execute huge routing processes, energy consumption increases if these routes have not been properly updated. This is why it is also important to assess route delays; they are directly related to the number of jumps that a node takes to reach a destination.

This work seeks to further advance the study of performance metrics in the analysis of sensor networks. One of the most important contributions of this work is the relationship between metrics and their influence on energy consumption in a WSN, as well as the comparison of how these metrics perform in widely known routing protocols with respect to a protocol proposed by the authors. An important assessment of these parameters is that routing tables provide critical information about network reliability and the number of valid routes available to the nodes for sending information. These metrics will directly influence packet loss rates in the network as well as information delivery reliability. The aforementioned study is implemented in a sensor network for each routing protocol in which the proposed energy model is also implemented, as described in Table 1. This model is based on the operation of the Texas Instruments CC2530 sensor [35] in active mode and stable conditions. Besides, for nodes under LoRa technology and WiFi, we use the model described in [36].

Table 1. Energy model base for each of the main tasks in a node [37].

	Voltage (mV)	Current (mA)	Time (ms)
Turn on mode	120	12	0.2
Microcontroller 32-MHz clock	75	7.5	1.7
CSMA/CA algorithm	270	27	1.068
Switching from reception to transmission	140	14	0.2
Switching from transmission to reception	250	25	0.2
Radius in reception mode	250	25	4.1915
Radius in transmission mode	320	32	0.58
Turn off mode	75	7.5	2.5

Global parameters denote the energy rates that each node spends for all activities performed within the network. These types of energy include the energy consumed by the microcontroller, which is regulated by the time the sensor remains on; the CSMA energy used when the node is listening to the channel to determine whether packets may be transmitted or received and the node is executing the CSMA/CA algorithm; the switching energy consumed when switching between activities, i.e., changing from transmission to reception or vice versa; transmission and reception energy used to transmit or receive a packet, respectively; awakening energy used when the node turns on; and the shutdown energy consumed when the node turns off.

All of these energies will promptly report how nodes save or consume energy, as the case may be, together with the routing protocol. Each type of energy is provided as an overall network metric. In other words, each energy is incorporated to observe possible points of interest or zoning in the network.

Local energy balance depends on the proximity of a node to the destination node. In the scope of this work, the final destination is the coordinator or collector node. As a local aspect, the energy consumption at each node is calculated, which denotes consumption depending on the distance between the node and the coordinator. Thus, the nodes closest to the collector will send their packets and also retransmit packets coming from other nodes. For this reason, the relational position of a node within the network significantly matters.

2. Description of the Energy Model

We propose a simple energy model under the basic tasks or activities that nodes perform in the network [38]. Figure 2 conceptually describes this scheme. This model considers all the energy components that contribute to the overall energy consumption under active mode. First of all, a node remains at start time (t_{ON}) to turn on. Then, it takes a switching time ($t_{Switching}$) to change status before sending a packet to the medium. Here, the node first runs the CSMA algorithm using a CSMA time (t_{CSMA}). Next, the node transmits an information packet expending a transmission time (t_{TX}). Now, the node takes a switching time ($t_{Switching}$) to change activities, it remains inactive ($t_{Inactive}$) and changes task again. In addition, it takes a switching time ($t_{Switching}$) to start receiving information and reporting a reception time (t_{RX}). The node performs these activities as many times as it sends and receives information (messages) during the sampling period. Finally, the node turns off expending a shutdown time (t_{OFF}). All of this while the microcontroller remains in active mode. This process measures the energy required for each main node activity in the network. Depending on the task and the time a node takes to execute it, this corresponds to a given voltage and current, so that the total energy used by each node can be obtained for each of the activities run for the network based on the previous model [37].

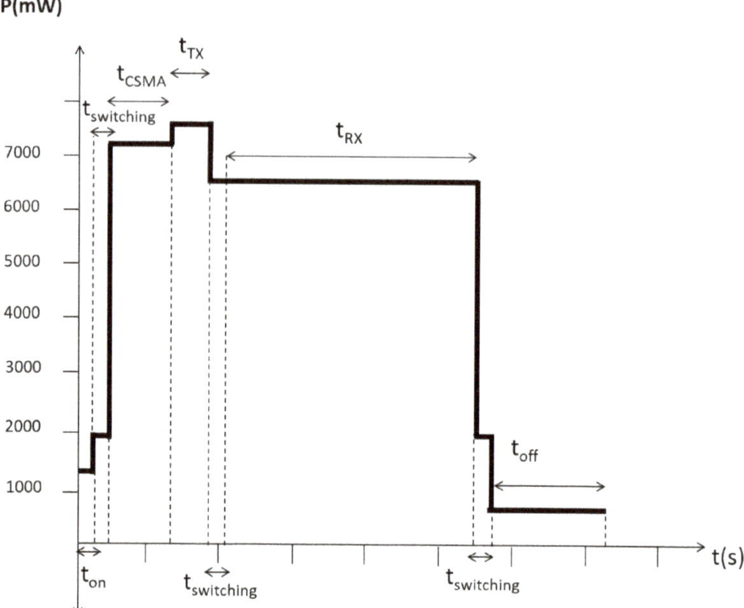

Figure 2. Conceptual energy scheme.

To compare the performance of the energy model, we used two types of sensors: CC2530 and CC2650 from Texas Instruments. We have coded the proposed model in a wireless sensor simulator programmed in C ++ based on the previously tested events [10]. The wireless module with antenna CC2530 nodes have 256 KB Flash and 8 KB of RAM sufficient for the implementation of widely known protocols in WSN. They have an IEEE 802.15.4 RF transmitter/receiver in 2.4 GHz high sensitivity (102 dBm). These sensors have four modes of operation in terms of energy savings and 21 general purpose pins, which cover the needs for connection of extra sensors. The CC2650 nodes are ultra-low power 2.4 GHz devices. An active and very low MCU and RF current and a low power mode current consumption offer excellent battery life. These devices contain a 32-bit ARM Cortex-M3 processor that operates at 48 MHz as the main processor and a set of extensive peripheral functions including an ultra-low power single sensor controller ideal for interconnecting external sensors.

Table 2 presents some results to verify the accuracy of the proposed energy model in predicting the total energy used by each node. We have taken as a basis the consumption for each type of energy proposed in the model. The described scenario is a configuration of 25 randomly and evenly distributed nodes and one of them is the coordinating or concentrating node. We left the nodes running from Monday to Sunday (seven days) on a university campus in an area of 500 × 500 m^2 under the conditions described in Table 3. We observed that the model described in Table 1 can be a reliable parameter to measure the general consumption of a wireless network of small to medium sensor size if the nodes are in active mode and the types of energies described give a faithful ideal of the main functions of a node in a network, taking approximately 4 to 5% margin of measurement error.

In this paper, an analytical model is proposed and exemplified by the operation parameters of the Texas Instruments CC2530 chip [35], which has a radio interface as per the IEEE 802.15.4 standard. This model shows the energy expended for each of the activities performed from the moment a node is added to a network and becomes part of it, listens to the channel, receives and sends messages, executes the link layer algorithms data, and changes states, ending with the energy consumed when it turns off and disconnects from the network.

Table 2. Energy model tested on two scenarios.

Scenario	E_{MC} (J)	E_{ON} (J)	E_{OFF} (J)	E_{CSMA} (J)	$E_{Switching}$ (J)	E_{TX} (J)	E_{RX} (J)
Simulator for CC2530	9.6398	2.9342	14.1883	78.4582	7.4382	59.9173	264.1783
Real network for CC2530	9.2578	2.5573	14.0123	78.9362	7.8625	59.1173	264.8746
Simulator for CC2650	8.4677	2.3891	12.1903	77.3274	6.1248	56.4702	260.9471
Real network for CC2650	8.4677	2.6891	12.1903	77.3274	6.1248	56.4702	260.9471

The energy used by the microcontroller depends on the node's operation mode. For example, techniques for turning nodes off reduce energy consumption by setting the microcontroller in idle mode for certain time intervals [39]. However, for this analysis, it is assumed that the mentioned SoC at each node operates in continuous active mode at 32 MHz (the microcontroller's clock frequency) to better study how energy consumption behaves under a specific routing protocol without the influence of techniques used for turning nodes off. Thus, the total energy used by the microcontroller will be given by (1):

$$E_{MC} = T_{MC} \times I_{MC} \times V_{MC}, \tag{1}$$

where T_{MC} is the time (seconds) taken by the microcontroller unit to consume V_{MC} (Volts) and I_{MC} (Amperes).

The starting energy is estimated based on the voltage, current, and time nodes required to turn on and be ready for the network, as described in (2):

$$E_{ON} = T_{ON} \times I_{ON} \times V_{ON}, \tag{2}$$

where T_{ON} is the time (seconds) that it takes to turn a node on and is given by V_{ON} (Volts) and I_{ON} (Amperes).

The model also describes the energy consumed when nodes turn off when the network time has ended (the sampling period). This energy is given by (3) and it is called shutdown energy.

$$E_{OFF} = T_{OFF} \times I_{OFF} \times V_{OFF}, \tag{3}$$

where T_{OFF} is the time (seconds) that it takes to turn a node off and is given by V_{OFF} (Volts) and I_{OFF} (Amperes).

Switching energy is expended when the node changes from receiving to transmission mode or vice versa. It is given by (4):

$$E_{Switching} = T_{Switching} \times I_{Switching} \times V_{Switching}, \tag{4}$$

where $T_{Switching}$ is the time (seconds) that it takes a node to change from the reception mode to the transmission mode or vice versa and it is given by $V_{Switching}$ (Volts) and $I_{Switching}$ (Amperes).

The CSMA/CA algorithm states that each time a node intends to transmit, it first checks if the channel is free from other transmissions. If it is, then the node proceeds to transmit. Otherwise, it calculates a random waiting time before attempting to listen to the channel again. The energy consumed by the CSMA/CA algorithm is given by (5):

$$E_{CSMA} = T_{CSMA} \times I_{CSMA} \times V_{CSMA}, \tag{5}$$

where T_{CSMA} is the time (seconds) in which a node computes the CSMA/CA algorithm and is given by V_{CSMA} (Volts) and I_{CSMA} (Amperes).

Transmission energy depends directly on the distance and interference. In this case, the nodes transmit their packets or forward packets from other nodes. Transmission energy is described as (6):

$$E_{TX} = P_{Length} \times T_{TX} \times I_{TX} \times V_{TX}, \tag{6}$$

where P_{Length} is the length of the packet (bytes), T_{TX} is the time (seconds) that it takes a node to send a byte and is given by V_{TX} (Volts) and I_{TX} (Amperes).

As with the transmission mode, a node expends receiving energy when it receives packets. This energy is determined using (7):

$$E_{RX} = P_{Length} \times T_{RX} \times I_{RX} \times V_{RX}. \tag{7}$$

where P_{Length} is the length of the packet (bytes), T_{RX} is the time (seconds) that it takes a node to receive a byte and is given by V_{RX} (Volts) and I_{RX} (Amperes).

The values of (1)–(7) are presented in Table 1.

Therefore, the total energy can be calculated with (8):

$$E_{Total} = E_{MC} + E_{ON} + E_{OFF} + E_{CSMA} + E_{Switching} + E_{TX} + E_{RX}. \tag{8}$$

The simple energy model shown by the aforementioned equations estimates the energy expended by the main tasks of a node within a WSN for any system, where those functions related to wireless communications are the most important from an energy standpoint.

This model can assess global and local energies. A generic case is presented to interpret the analysis of the model. The energy of the node i is represented as $EnergyNode_i$, which is obtained by adding up the energy consumed by a node i to perform each task along the active network. When a node connects to the network, it has consumed zero energy initially, so that $EnergyNode_i = 0$.

For this energy model, we segregate energies that depend directly on the number of packets transmitted or not. The energies related to the packets are: E_{TX}, E_{RX}, $E_{Switching}$ and E_{CSMA}. The energies that depend exclusively on node operation are: E_{ON}, E_{OFF}, and E_{MC}. Therefore,

$$EnergyNode_i = \overbrace{E_{TX_i} + E_{RX_i} + E_{Switching_i} + E_{CSMA_i}}^{\textit{Dependent on packets}} + \overbrace{E_{MC_i} + E_{ON_i} + E_{OFF_i}}^{\textit{Independent on packets}}. \tag{9}$$

Now, to calculate the energy consumed by each node starting with packet-dependent energies, E_{TX_i} will be the energy transmitted in the node i, so,

$$E_{TX_i} = (P_{Length} \times T_{TX} \times I_{TX} \times V_{TX}) \times (P_{TX_i} + P_{RTX_i}), \tag{10}$$

where P_{TX_i} is the total number of messages transmitted by node i. P_{RTX_i} is the total number of packets retransmitted by node i, because a packet receipts acknowledgment, ACK, was not received.

In addition, E_{RX_i} will be the energy transmitted in the node i, so,

$$E_{RX_i} = (P_{Length} \times T_{RX} \times I_{RX} \times V_{RX}) \times P_{RX_i}, \tag{11}$$

where P_{RX_i} is the total number of packets received by node i.

In the same way, $E_{Switching_i}$ is the switching energy consumed when a node changes from transmission to reception or vice versa. The $E_{Switching_i}$ energy for node i is given by (12):

$$E_{Switching_i} = (T_{Switching} \times I_{Switching} \times V_{Switching}) \times (P_{TX_i} + P_{RTX_i} + P_{RX_i}), \tag{12}$$

The term $(P_{TX_i} + P_{RTX_i} + P_{RX_i})$ represents the number of times there is switching or changing in state activity from the time the packet is sent to the time the packet is received or vice versa. Therefore, the best option is when a node sends a packet to a channel that is free and an acknowledgment ACK is received. Then, $P_{TX_i} = 1$ and $P_{RX_i} = 1$. If the node receives the ACK, it does not retransmit the packet and $P_{RTX_i} = 0$. This switching operation is performed in two different stages: the node uses switching energy to change state, transmits the information, uses again switching energy, and receives the packet information.

Transmission and receiving energies are considered by (10) and (11), respectively.

For each packet transmitted, the node i executes the CSMA/CA algorithm. The corresponding energy consumed, E_{CSMA_i} is given by (13):

$$E_{CSMA_i} = \begin{aligned} &(T_{CSMA} \times I_{CSMA} \times V_{CSMA}) \times \\ &(P_{TX_i} + P_{RTX_i} + N_{RT_i}), \end{aligned} \tag{13}$$

where N_{RT_i} is the number of times that the CSMA/CA algorithm goes back to calculate a delay from the time the channel was found to be busy. This is a random variable. That is, it takes a different value for each transmission. The $(P_{TX_i} + P_{RTX_i} + N_{RT_i})$ term implies that before each message transmission, the channel auditing process must be executed (CSMA/CA algorithm). In the best case, $P_{TX_i} = 1$ in the $(P_{TX_i} + P_{RTX_i} + N_{RT_i})$ term, because the channel was free. Then $N_{RT_i} = 0$, (there were no retries listening to the channel, i.e., CSMA retries), and the packet is transmitted successfully with $P_{RTX_i} = 0$ (without retransmissions). However, the network conditions are not always ideal and there will be collisions causing packet retransmissions and retries when listening to the channel. Then, the N_{RT_i} and P_{RTX_i} variables will exhibit non-zero values.

Throughout the sampling time ($T_{Sampling}$), the node consumes energy from the microcontroller in active mode for this model. Therefore,

$$E_{MC_i} = T_{Sampling} \times I_{MC} \times V_{MC}. \tag{14}$$

Now, when the sampling time begins, all nodes turn on. Then, the initial energy consumed by the node i is $E_{ON_i} = T_{ON} \times I_{ON} \times V_{ON}$. Finally, when the time has finished, node i turns off, consuming shutdown energy, so $E_{OFF_i} = (T_{OFF} \times I_{OFF} \times V_{OFF})$.

At the end of the sampling period, the node reports the total energy consumed by all the functions it performed during the network processes, which will be the local energy of the node. Thus,

$$EnergyNode_i = \begin{aligned} &E_{ON_i} + E_{MC_i} + E_{OFF_i} + \\ &E_{Switching_i} + E_{CSMA_i} + E_{TX_i} + E_{RX_i}. \end{aligned} \tag{15}$$

To obtain the total node energy to assess the global energy of the network, we add the total energy of each node when the sampling time has ended as shown in (16):

$$Total_Energy = \sum_{i=1}^{Total_{Nodes}} EnergyNode_i, \tag{16}$$

where $totalNodes$ is the total number of nodes.

Equation (15) describes the actions performed by node i since it turns on (consuming initial energy E_{ON_i}), and becomes part of a network, until the moment it turns off. The most complex stage is when the node listens to the channel. This activity causes the node to decide whether to transmit packets. All packets sent (either transmitted or retransmitted) imply that the node must first listen to the channel. That is, the node executed the CSMA/CA algorithm using CSMA energy (E_{CSMA_i}). If the channel is free, the node consumes transmission energy (E_{TX_i}) to transmit the packet (P_{TX_i}). If the channel is busy, the node generates a retry variable (N_{RT_i}) because it has to execute the CSMA/CA algorithm again. After the node sends the packet, it receives an acknowledgment, ACK, using reception energy (E_{RX_i}) and receives the packet (P_{RX_i}). On the contrary, if the node does not receive an ACK, it retransmits the packet generated as a retransmission (P_{RTX_i}). For all activity changes, the node consumes switching energy ($E_{Switching_i}$). After this basic single packet cycle, the node continues to receive, listen, transmit, and changes states until the simulation time has ended and the node turns off using shutdown energy (E_{OFF_i}). During all the processes performed by a node as part of a network, the node is in active mode and the microcontroller remains active throughout the sampling period, consuming microcontroller energy (E_{MC_i}).

Algorithm 1 shows the model operation pseudo-code for each of the energies previously exposed in the main processes executed by a node within the network.

Algorithm 1: Algorithm for the proposed energy model.

```
E_{ON-Total} = 0
E_{MC-Total} = 0
E_{OFF-Total} = 0
E_{Switching-Total} = 0
E_{CSMA/CA-Total} = 0
E_{TX-Total} = 0
E_{RX-Total} = 0
FOR{Every node i}
Node i turns on with a coverage radius;
E_{ON-Total} = E_{ON-Total} + E_{ON_i}
E_{MC-Total} = E_{MC-Total} + E_{MC_i}
Node i listens to the channel;
E_{CSMA/CA-Total} = E_{CSMA/CA-Total} + E_{CSMA/CA_i}
E_{MC-Total} = E_{MC-Total} + E_{MC_i}
IF{channel is free }
Send neighbor acknowledgment packet;
E_{TX-Total} = E_{TX-Total} + E_{TX_i}
E_{RX-Total} = E_{RX-Total} + E_{RX_i}
E_{MC-Total} = E_{MC-Total} + E_{MC_i}
ELSE
Increase delay and re-check channel;
E_{MC-Total} = E_{MC-Total} + E_{MC_i}
END IF
Node i populates its neighbors table;
IF{Node i requires sending information}
Find a route in neighbors table;
IF{channel is free}
Send neighbor acknowledgment packet;
E_{TX-Total} = E_{TX-Total} + E_{TX_i}
E_{RX-Total} = E_{RX-Total} + E_{RX_i}
E_{MC-Total} = E_{MC-Total} + E_{MC_i}
ELSE
Increase delay and re-check channel;
E_{MC-Total} = E_{MC-Total} + E_{MC_i}
END IF
END IF
IF{Simulation Time == end}
E_{OFF-Total} = E_{OFF-Total} + E_{OFF_i}
E_{MC-Total} = E_{MC-Total} + E_{MC_i}
END IF
END FOR
```

2.1. Limitations of the Energy Model

This subsection describes some of the limitations of the proposed energy model. This model comprises the main types of energy required for a node to perform generic functions within the network. However, this model does not describe specific energies related to the anomalous behavior of the nodes, yet it is capable of predicting strange behaviors in the network due to the alteration of typical node functions. These behaviors are due to situations of stress, such as node connections and disconnections due to channel intermittence. Likewise, these situations can be represented as a result of other types of energy. For example, in network areas with intermittent nodes or links, nodes frequently disconnect. This situation reflects in increased retransmissions, channel auditing retries, and packet collisions. These changes increase the transmission, reception, switching, and energies of the CSMA/CA algorithm. Thus, the proposed model can identify anomalous behavior observed in these types of energy and draw conclusions regarding risk areas or potential attacks.

The main function of a WSN is sending information to one or more collector nodes and detecting the conditions of the channel to find the optimal route. Then, in these networks, it is possible to implement different sleeping node techniques. That is to say, there are periods of time in which a certain number of nodes enter into passive mode, consuming less energy than they normally do. In this study, we do not look into sleeping node energy. However, the model can be extended, adding this energy equation, if the voltage and current of a sensor in a given time interval are available. For these reasons, the proposed model is simple, easy to implement, and quite scalable as per analysis requirements.

3. Analysis and Results under the Simulation Tool

One of the objectives of this work is to assess critical metrics in WSNs. In this section, the performance of the MPH routing protocol is compared against three sensor network protocols: AODV, DSR, and ZTR. The performance metrics used for comparison are the total energy consumption, delay, overload, resilience, valid routes, and number of jumps. This analysis is performed to analyze the influence of these parameters on energy consumption.

This is used to assess possible anomalies in a network or an area at risk of attack. The delay and energy consumption in a network are directly related to the complexity of the routing algorithms. When a routing protocol uses many algorithms and processes to send a packet, the sensor nodes will introduce long delays and incur in high-energy consumption. The MPH routing protocol is not only characterized by its hierarchical topology, but also by the origin touting and fast reconfiguration of its topology when faced to an unexpected change. This has to do with the resilience of the network (the network's ability to recover from unexpected changes).

A routing protocol provides reliability when nodes exhibit in their tables (either routing or neighbor tables) valid routes, i.e., routes that have not expired or that have not been invalidated by a node disconnection. Overloading the routing protocol will affect the channel's occupation because it is constituted by the control packets that a routing protocol needs for the reliable delivery of information. The lower the overload, the lower the probability of collisions and, in turn, the lower the number of packet retransmissions. A small overhead ensures that the information will be delivered more quickly and reliably.

For simulations, a grid layout with 22 nodes described in Figure 3 is considered. The nodes are located randomly in an area of 300 m × 300 m and nodes have a coverage range of 15 m. A data rate of 250 kbps and a packet length of 22 bytes were taken. Table 3 shows the simulation parameters. Each node contributes a traffic of 10 packets per second. There are two coordinator nodes marked in red color. All network nodes can send packets to the coordinator nodes directly (one hop) or indirectly (multiple hops that form a route). There are nodes near to the coordinator and all other nodes must send packets through these nodes, causing them to drain their batteries faster because they constitute the network bottlenecks.

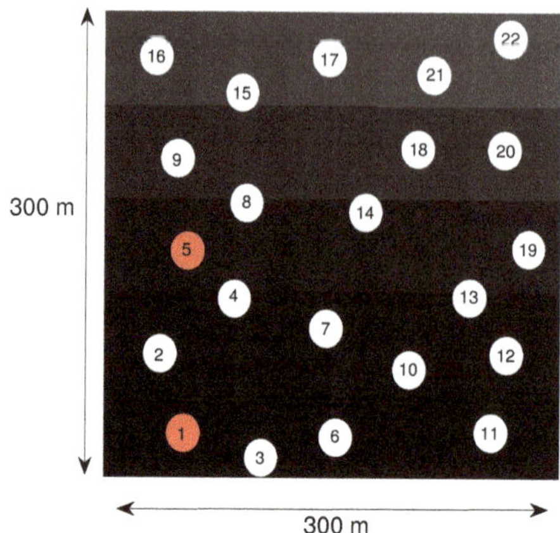

Figure 3. The network topology considered.

Table 3. Simulation and real network parameters under CSMA/CA carrier sense multiple access with collision avoidance [34].

Parameter	Value
Physical Layer Parameters	
Sensitivity threshold receiver	$-94\,\mathrm{dBm}$
Transmission power	$4.5\,\mathrm{dBm}$
MAC Layer Parameters	
Maximum retransmission number	3
Maximum retry number	5
Maximum number of tries to reach a node from the collector	9
Packet error rate	1%
Average frame length	22 bytes
Maximum number of backoffs	4
MAC protocol	IEEE 802.15.4
MAC layer	CSMA/CA
Network Layer Parameters	
Number of nodes	22
Maximum data rate	250 kbps
Scenario	Static nodes

Figures 4–9 show, for some well-known protocols in the literature, the energy types per crown for the distribution shown in Figure 3. These graphs show the numerical analysis for each type of energy described in the proposed energy model. A 24 h/7 day execution of 22 sensors was carried out in an area of $300 \times 300\,\mathrm{m}^2$ of the university campus of Universidad Panamericana in Guadalajara, Mexico. A set of high-level wireless communication nodes CC2650 and CC2530, based on the IEEE 802.15.4 standard, were used. The space covered by the sensors is an area with trees, buildings, traffic of people and vehicles with both indoor and outdoor spaces. The distribution structure of the sensors is the one shown in Figure 3. We aimed at taking six representative protocols in WSN such as: AODV, DSR, LEACH, PEGASIS, MPH (proposed protocol), and ZTR. Crowns were formed by the proximity of the nodes with respect to the coordinating node. A crown has the characteristic that the nodes belonging to it have similarity with respect to their performance parameters, i.e., with approximately the same distance to the coordinating node, the nodes have more or less the same traffic and forwarding packets

to their destiny. This is why the nodes are grouped into categories called crowns, which make them take a specific level of similarity in the network. For these graph results, the network will only have a coordinating node and this will be node 1. It is important to note that crowns are formed based on the amount of links that each routing protocol forms; not all protocols have the same number of crowns to analyze. The proposed energy model lets us know exactly how much energy an average node spends on each crown. The energy model is applied to the programming and operation of the sensors to establish their energy consumption separately, depending on the type of energy. As the model is validated, it allows global and local knowledge of the energy of the nodes according to the performance parameters such as: the proximity to the coordinating node, the number of established links, the collisions that generate the amount of traffic and control packets, the delay in the delivery of information, and the processing of resources at the node level.

In Figure 4, the AODV protocol presents six crowns of nodes. This may be because AODV is a reactive protocol that forms links in all directions, so the network becomes a mesh with widely redundant links. However, the fact of having so many links may produce an increase in packet collisions and protocol control packets, which generates more packet retransmissions and listening attempts to the communication channel. The advantage of presenting several crowns or levels of nodes with similar performance is the redundancy and the amount of different routes that a package can take to reach its destination. The disadvantage presented by this number of packets administered by the routing protocol is the high-energy consumption due to the possible loss of packets and their retransmission. In AODV, the crowns with the highest energy load are 1 and 2, which are the closest nodes to the coordinating node. This is understandable because they are the nodes that forward the traffic of the other nodes in the network and the coordinating node generates a bottleneck. We observe that between the last crown (crown 6) and the first two crowns, the difference in energy expenditure is 48%, which shows that nodes farther from the coordinating node have less traffic load, less collisions and less retries listening to the channel, then the CSMA algorithm runs in less times. Due to the energy model, we can also note that in crowns 1, 2, 3 and 4 the transmission and CSMA energies are similar. This may be due to the strong weight of the network that is located at the center of the topology; the AODV protocol creates a mesh and not a link tree.

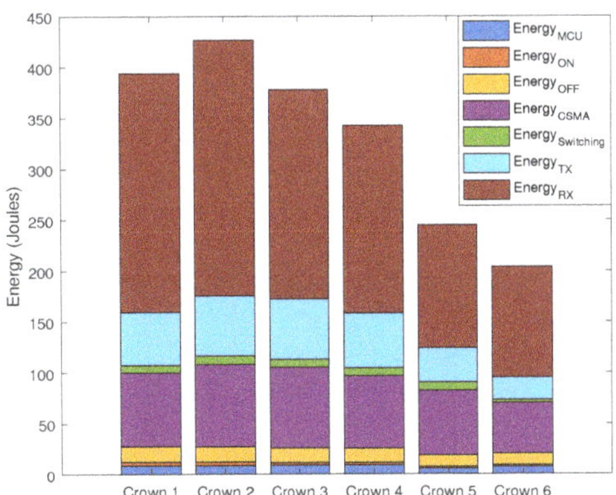

Figure 4. Energy types per crown for the AODV protocol.

In Figure 5, the DSR protocol presents a structure and a quantity of crowns similar to AODV. These two protocols are the highest energy consumers of all the protocols studied in this work, with a 15% higher expenditure of energy in the nodes. Crowns 1 and 2 have the highest energy expenditure due to the redundancy of the protocol links. Although in DSR the links form a mesh in the network and there is a large amount of packet flow, this protocol, unlike AODV, has a more marked energy expenditure per crown; the first being the most consumed, and the last, the one that consume the less. For DSR, the energy difference between the first and the last crown is 42%. In addition, due to the topology configuration that DSR generates, there are many surrounding packets in the network (traffic and control), which generate packet losses, retransmissions and therefore, listening retries to the communications channel to determine if it is already available or still in use. This can be noted with the fact that CSMA energy is similar in almost all crowns. The AODV and DSR protocols, being both reagents, have a similar energy expenditure with a difference of only 6%, even though in the last crown, the transmission energy decreases in DSR because the crowns are more scaled than in AODV.

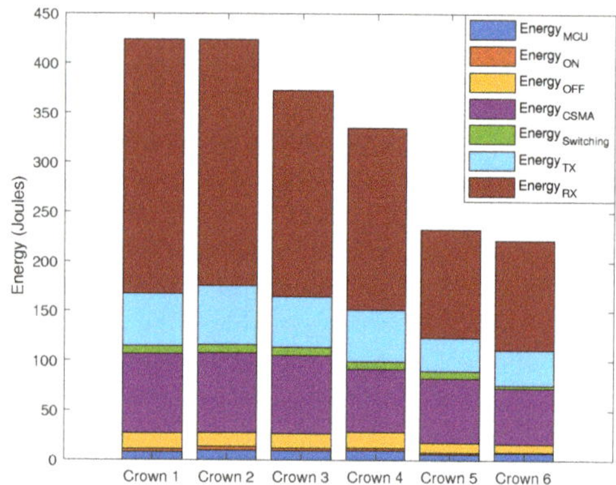

Figure 5. Energy types per crown for the DSR protocol.

In Figure 6, the LEACH protocol uses techniques to reduce collisions between clusters and within the clusters themselves. Data collection is centralized and runs periodically, which is a characteristic of a proactive protocol. The protocol configuration for this scenario points that for one third of each day, a node near the lower left corner of the topology (near node 1) will be the coordinating node, for another third of the day, it will be a node in the middle of the topology, and finally, for the remaining third, it will be a node near the top right corner of the topology (near node 22). Due to the imbalance that is established in these changes of roles of the nodes, there is less crowns in the network and almost all have approximately the same energy expenditure. The above allows stating that all nodes belong to the same crown. LEACH assumes that all nodes transmit with sufficient power to reach the coordinating node and that each node has sufficient computing power to support different MAC protocols. In practice, this is complicated and, as it can be seen in real cases, the first crown differs energetically from the last one by 6%. However, the existence of cluster and various roles of the nodes, allows reducing energy by 13% with respect to reactive protocols such as AODV and DSR. In LEACH, the transmission energy is almost the same in all crowns, except for the last one, with a difference of 5% with respect to the others.

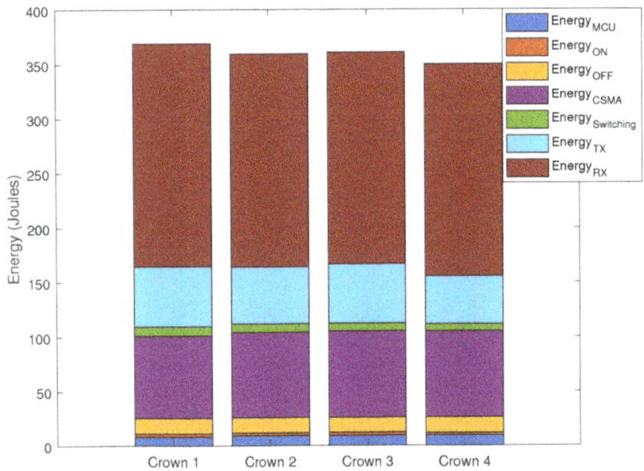

Figure 6. Energy types per crown for the LEACH protocol.

In Figure 7, the PEGASIS protocol extends the life of the network by limiting the nodes communication only to their closest neighbors and take turns communicating with the coordinating node. This protocol uses networking techniques and allows only local traffic between nodes that belong to the same region or crown to reduce bandwidth consumption. One of the great advantages of PEGASIS is that the distance between the nodes is calculated based on the intensity of the signal; the links are really strong, thus preventing packet retransmissions. We observe that the difference in energy consumption between LEACH and PEGASIS is 14%, with PEGASIS demonstrating the greatest savings. Energy improvement occurs by avoiding overload caused by LEACH's dynamic generation of the cluster and by minimizing the number of packet transmissions and receptions using the data aggregation technique. PEGASIS assumes that each node must be able to communicate with the coordinating node directly and that each node contains a complete database of the location of the other nodes in the network. This reduces network performance a bit by making processing slightly heavier. The energy of the crowns is similar, only with a difference between them of 5% and due to the role relay in the nodes; the crowns are not scaled from higher to lower consumption.

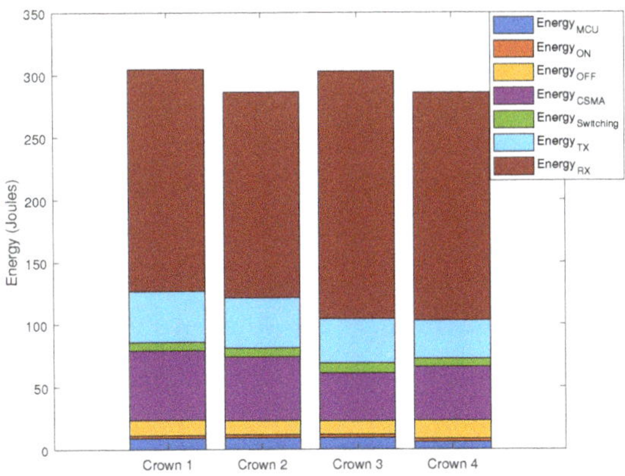

Figure 7. Energy types per crown for the PEGASIS protocol.

In Figure 8, the MPH protocol is a hybrid protocol (predominantly proactive). It establishes link hierarchies based on the proximity of a node to the coordinating node. This hierarchical tree topology only allows a few links, but there is a sufficient degree of redundancy. In MPH, we observe five crowns scaled around the coordinating node with an energy difference of 10% between the first and the last one with a similar transmission/reception energies in all crowns. MPH takes advantage of the fact that there are no links among nodes of the same hierarchy level, which decreases the cost in processing the neighbor tables and decreases the amount of protocol control packets. The difference in energy consumption between AODV and DSR with respect to MPH is 40% in favor of MPH.

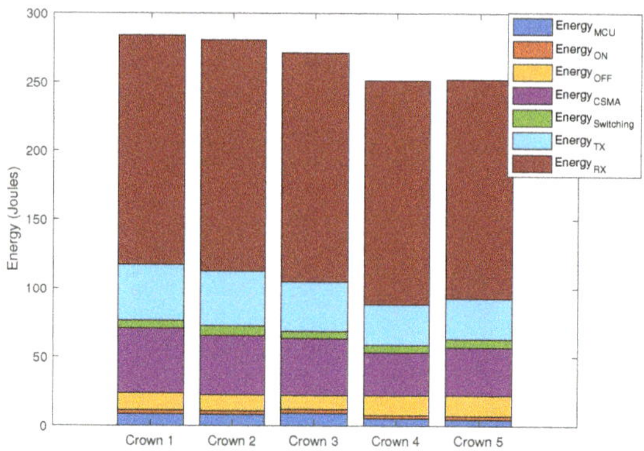

Figure 8. Energy types per crown for the MPH protocol.

The ZTR protocol, shown in Figure 9, is a proactive protocol, simple, and easy to implement. It consists of an algorithm with limited resources that performs multi-hop routing without route discovery procedures and is based on a hierarchical distribution scheme. As in MPH, we note that five scaled crowns are established from the highest to the lowest energy consumption. ZTR has a 5% energy saving with respect to MPH because its links are simpler and the nodes cannot have more than one parent node. The small difference in energy expenditure that is established between MPH and ZTR, being ZTR so simple, is because several packages can be lost due to the low redundancy of links but this fact is compensated with the ZTR's speed of information delivery.

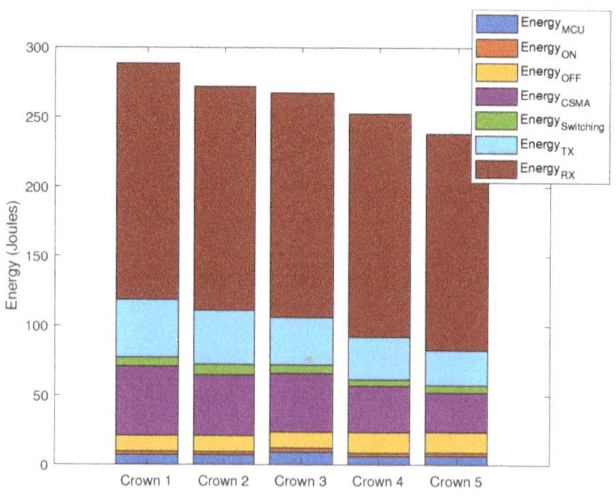

Figure 9. Energy types per crown for the ZTR protocol.

Now, having tested the energy model for the random scenario, Figure 10 shows the energy distribution of the nodes for each observed protocol: AODV, DSR, LEACH, PEGASIS, MPH, and ZTR. This scenario is based on the topology presented in Figure 3. The packet sending rate is 10 packets per second and the links have a packet loss between 0.5 and 1.5%, measured on a one day window with an average per hour for each node. According to the results, it can be observed that approximately 75% of the nodes in MPH and ZTR present values below 0.10 Joules while for AODV, 75% of the nodes distribute their energy in values between 0.14 and 0.19 Joules. Regarding DSR, the energy values for the 75% of the nodes are distributed between 0.16 and 0.19 Joules. The MPH and ZTR protocols present some extreme values that show their proactive nature, in which the creation of hierarchical routes and the amount of energy can be concentrated in the nodes near to the collector nodes. We can also note that the MPH protocol has a more compact energy distribution between randomly distributed nodes, even more compact than ZTR since both have lower energy consumption compared to AODV and DSR. Compared to DSR, AODV has a marked distribution of energy between nodes; this indicates that the crowns, which nodes forward packets and are around the coordinator nodes, have more significant differences in energy consumption in AODV than in DSR. LEACH and PEGASIS present the most compact distribution of energy in the network nodes. The values of 50% of the nodes range between 0.06 and 0.078 Joules for LEACH and their most extreme value falls to 0.042 Joules, this can occur in the crowns furthest from the coordinating node. In this protocol, the farthest nodes from the coordinating node can lead to a shorter survival time and generate greater packet delays. For PEGASIS, 50% of the nodes have energy values between 0.06 and 0.07 Joules and have an outlier at 0.041 Joules. Due to the propagation of chain packets, this protocol is the most efficient in energy consumption. This protocol reduces both the bandwidth requirement and the overhead.

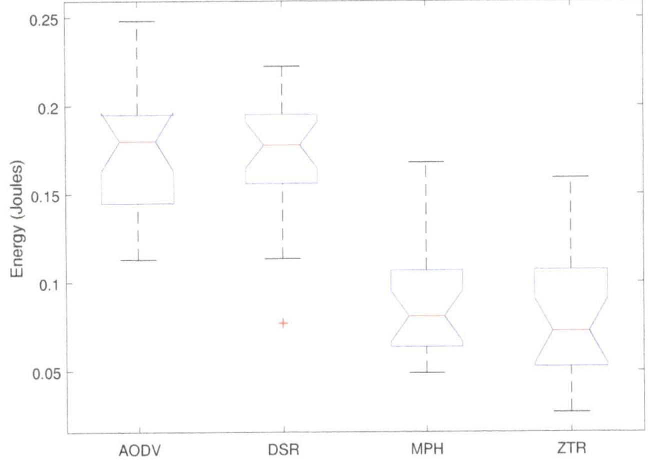

Figure 10. Energy distribution in the nodes for each evaluated protocol.

In Figure 11, energy transmission is varied in order to increase the coverage radius. Its aim is to show how energy behaves in each of the studied protocols. These conditions can also evaluate some type of resilience in the network according to each protocol. The fact of increasing the coverage radius generates more collisions in the network. Although the routes can be compensated, because they are shorter, nodes have a greater number of directly connected neighbors. Even so, the energy model shows that the MPH protocol has similar characteristics to ZTR, the latter being extremely simple and not scalable but quite fast and efficient. MPH shows an average energy saving of 10% with respect to ZTR, 24% with respect to DSR, and 28% compared to AODV.

In addition, we observe that LEACH and PEGASIS have similar behaviors with a difference of 3% between them. These are the protocols that have the lower energy consumption in face of the stress caused in the network. When the transmission power is increased, these protocols maintain the rotation of roles in the nodes and therefore, both the energy and the network overload are balanced. The above is perfectly combined with the fact that both are hierarchical protocols further enabling packet traffic. PEGASIS presents a decrease in the total energy for each radio due to its approach of sending packets in chain, unlike the establishment of clusters exhibited in LEACH. MPH differs from LEACH and PEGASIS by 30%. This may be due to the proactive nature of MPH, in which from time to time, the tables of neighbors are updated and generate greater overhead. If the radius increases, the tables of neighbors become bigger and their update more complex.

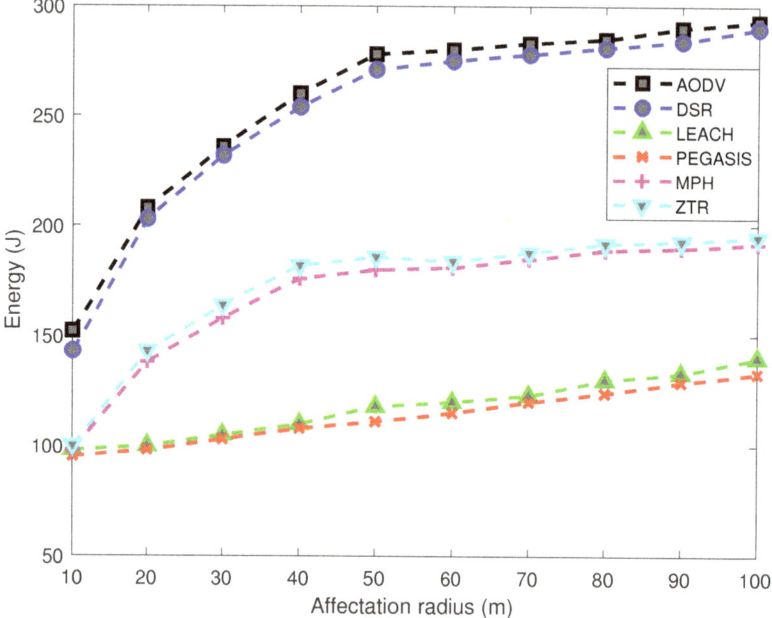

Figure 11. Energy according to the increase in the coverage radius of a node.

Figure 12 shows two comparative conditions for the six protocols under study: stable network conditions working properly and adverse network conditions under the topology referred in Figure 3. For the study, we took an average of energy at the same hour (noon) for 7 days for each of the nodes in the network. Nodes transmit at a rate of 70 packets per second. To generate interference and create adverse conditions in the network, we put 5 nodes close to the majority of nodes in the network that were emitting the same reactive jamming frequency, thus increasing the loss of packets along the links as it is shown in Figure 13 with unnumbered jammer nodes marked in blue. The network has a stress zone and a high focus on packet loss. This was made to analyze how the network reacts with each routing protocol and how this influences the energy of each node having both local and global perspectives regarding energy consumption of the network.

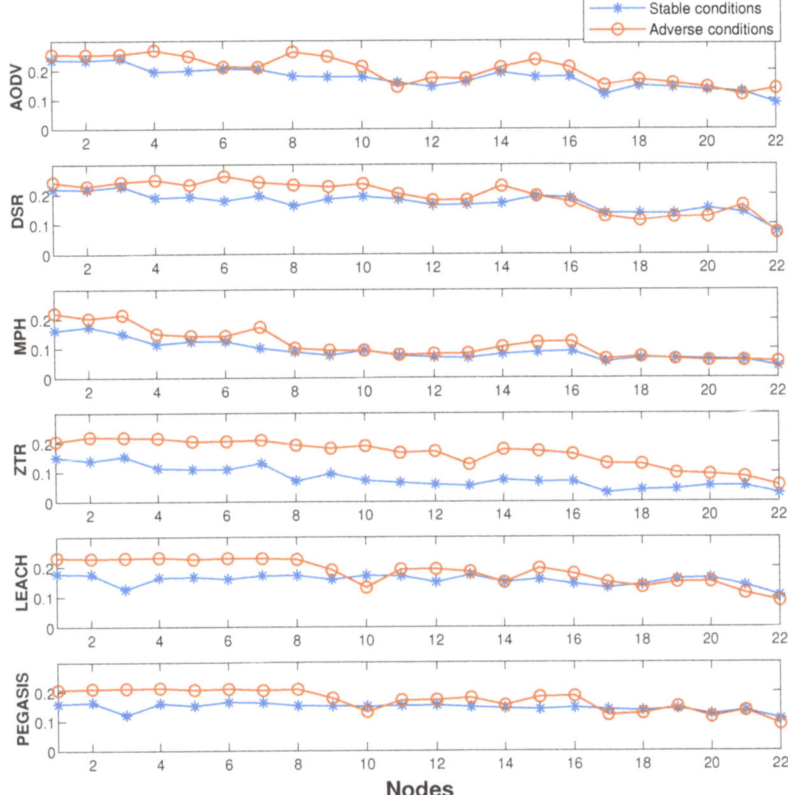

Figure 12. Energy at each node of the network under both stable and adverse conditions for the four protocols studied.

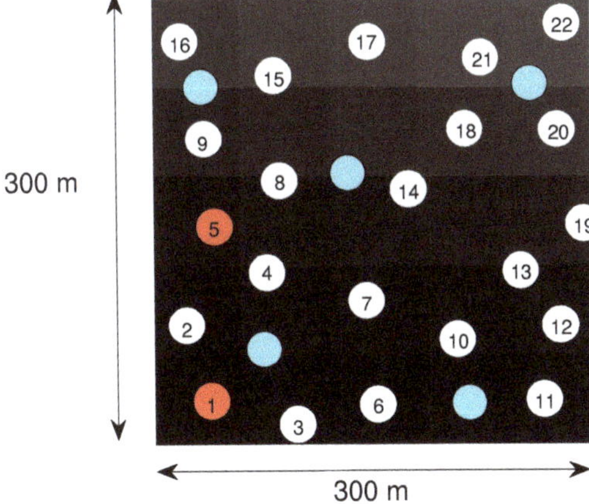

Figure 13. Grid topology with jammer nodes generating adverse conditions.

We observe that node energy in the surroundings of that half greatly increased in adverse conditions for AODV and DSR. ZTR and MPH try to stabilize the energy of the network nodes and redistribute the packet losses. LEACH and PEGASIS have a similar energetic behavior that does not necessarily behave by crowns from greater to lesser overload. These protocols react very well before adverse conditions because in more than 50% of the nodes, the energy expenditure is similar in both stable and adverse states. These protocols have a similar energy expenditure between them with a difference of 5% in favor of PEGASIS and a difference of only 3% with respect to MPH and ZTR. This is reflected in the fact that the average energy per node in adverse conditions for the network are 0.1990, 0.1968, 0.1805, 0.1741, 0.1671, and 0.1150 Joules for AODV, DSR, LEACH, PEGASIS, ZTR, and MPH, respectively. The above figures show that MPH has an energy expenditure similar to ZTR, which is a simple and fast algorithm; still MPH exhibits the advantage of route maintenance and redundancy. ZTR and MPH have similar natures, both having proactive characteristics. However, when there is a large number of packets in the network, there are more collisions and packet loss. The ZTR protocol can lose links and some isolated nodes might remain causing an increase in energy consumption. For this reason, the energy difference between MPH and ZTR is 31%. MPH, having multi-parent links, shows greater redundancy and increases the amount of valid routes that continue to function for changes in the network. This can be interpreted taking into account that MPH was designed to combine the best design features of a proactive protocol with the redundancy of the reactive protocols and the foregoing together with the verification of the proposed energy model, which allows visualizing easily and with concise data possible anomalies in specific areas of a network.

In Table 4 a sampling period of 100 seconds was taken for the simulations. Tests were performed every 10 seconds, which is the percentage of Packet Delivery Ratio (PDR). The PDR is a metric that indicates the number of packages delivered in a given time and indicates the collisions present in a network that cause packet loss. This performance metric is directly related to the energy and provides an idea on how nodes behave in the network; for example, if there are connections and disconnections of nodes, interference and quality of the links, among other features that show the performance of the routing protocol. As aforementioned, AODV and DSR have a similar behavior because their reactive nature. Initially, the network nodes do not know the routes to the destination, so they send request packets implying a large number of control packages, which increases the overhead. The LEACH and PEGASIS protocols try to make the route to the coordinating node efficiently through clusters or changes in the role of the coordinating nodes; this increases the assertiveness of the nodes in the package delivery, a consequence that is reflected in energy consumption. The difference in PDR in this initial stage between AODV and DSR with respect to LEACH and PEGASIS is approximately 23% in favor of the latter protocols. MPH and ZTR are proactive protocols, so at the initial stage, nodes behave the same as in stable state because the neighbor tables are periodically renewed and not when the routes require it. One of the big differences between MPH and ZTR is multi-parent links, which allow the packets to have greater redundancy in MPH and a reliable delivery. The difference in PDR between AODV and DSR with respect to MPH and ZTR is 28% in favor of MPH and ZTR.

Table 4. Percentage of Packet Delivery Ratio (PDR).

TIME (s)			%	PDR		
	AODV	**DSR**	**LEACH**	**PEGASIS**	**MPH**	**ZTR**
10	72	73	92	93	98	93
20	81	84	90	93	98	89
30	92	91	91	94	98	89
40	97	97	91	93	98	90
50	93	91	92	94	96	91
60	83	84	93	95	97	91
70	86	86	93	93	99	91
80	93	93	90	93	98	90
90	96	97	90	93	96	89
100	92	93	90	94	98	91

Concerning the rate of transmitted packets, we analyze the measurement impact of the energy model when we vary the packet transmission rate. Tests were carried out for 1 day (24 h). In particular, on a Wednesday; the day in which we can find regular traffic of people and vehicles on campus, and therefore, more operating wireless devices. We show these results in Table 5.

Table 5. Effect of packet transmission rate (PTR) variation on energy.

PTR (kbps)			Total	Energy (J)		
	AODV	**DSR**	**LEACH**	**PEGASIS**	**MPH**	**ZTR**
50	33.4	31.3	20.4	20.1	21.1	20.2
100	35.4	34.7	23.8	24.3	25.4	23.1
150	42.3	41.5	30.4	27.4	32.5	31.2
200	44.6	43.2	32.1	30.2	34.2	34.5
250	45.8	43.1	32.3	31.4	34.6	35.1

We note that, for five different packet transmission rates, the most drastic variation is found in the rates of 50 and 100 kbps, with a 37% difference with respect to energy at 250 kpbs. The difference in energy expenditure with the change in packet transmission rates is 43% higher for reactive protocols, compared to other protocols. The difference between proactive protocols and energy-aware protocols is only 5% in favor of the latter.

3.1. Experimental Scenario

To validate the tests done by the simulator, we intended to recreate a scenario with real sensors on the university campus of the Universidad Panamericana in Guadalajara, Mexico. We considered an area of 300 × 300 m^2 surrounded by a two-story building, a green area, living rooms, and free area with a terrace, as shown in Figure 14.

Figure 15 shows the top view of the engineering building at the university campus and the location of the sensors in this building. This plan helps to better visualize the walls and the specifications of the construction to give a better idea of the traffic routes of people, location of computer equipment, lights, etc.

There is a Gateway with LoRa technology and WiFi, which is the general hub of all the nodes. This Gateway receives data from all devices and sends the information to a platform called the things network. The devices that are held as nodes are divided into three technologies: Bluetooth, LoRa, and Zigbee.

Zigbee nodes are a set of high-level wireless communication protocols, based on the IEEE 802.15.4 standard, i.e., communication using the 2.4 GHz frequency. The CC2530 device is configured as the coordinator; its main function is to create the network in mesh topology. The other four devices can be connected to the network and have communication between them, allowing the possibility to have devices withdrawn from the coordinator because the information packet passes through the other devices to the coordinator. Each of the devices contains a temperature sensor, allowing measuring the ambient temperature in different locations of the university campus. The coordinator sends a signal to know which devices are connected to the network and in turn, these send their temperatures every 5 min. After obtaining all data, the coordinator sends the information through another protocol called LoRa to the Gateway.

Figure 14. The test area at Universidad Panamericana (Mexico).

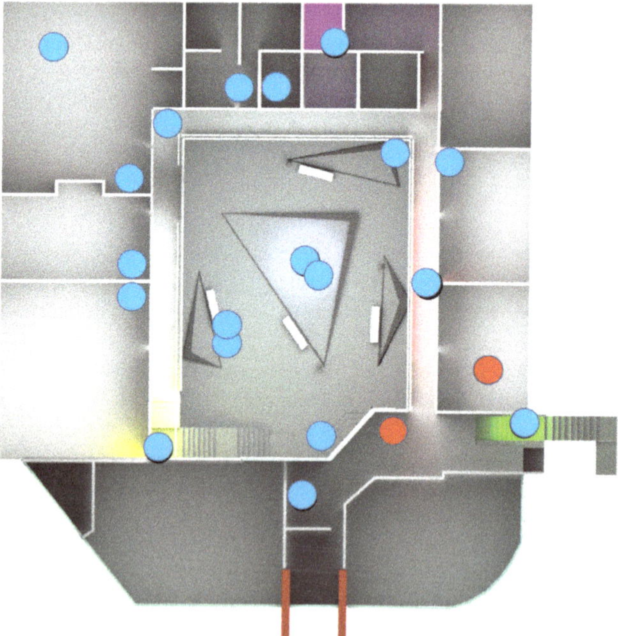

Figure 15. Top view of the test area (engineering building at Universidad Panamericana, Mexico).

For LoRa nodes, LPWAN is a specification for low power and wide area networks designed specifically for low power consumption devices operating in local, regional, national, or global networks. In Mexico, LoRa uses the 915 MHz frequency. The topology of LoRa is point-to-point; there is a gateway or hub and one or more nodes. The Gateway is in charge of reading all the packets that are on that frequency. The nodes are devices that transmit small information frames to avoid high-energy consumption.

The created network consists of five different devices: garbage sensor, light sensor, accelerometer sensor, gyroscope sensor, and environmental sensor. The Gateway is the device that acts as connection interface between devices and allows resource sharing between two or more computers. The Gateway used contains the LoRa and WiFi protocols. It obtains all the transmitted data through the LoRa protocol; data are then transmitted via WiFi allowing them to be found on the ThethingsNetwork platform. In this platform, it is possible to display the data separately from each node, knowing when the last transmission was made, the frequency of transmission of each device, and its measurements. This device obtains all the data transmitted by the Zigbee coordinator, all the data transmitted by the LoRa nodes, and also the data transmitted by the Bluetooth concentrator.

For Bluetooth nodes, Wireless Personal Area Networks (WPAN) is an industrial specification operating on the frequency of 2.4 GHz as well as Zigbee. There are Bluetooth nodes that work as beacons (low power consumption devices that emit a broadcast signal). In this case, each node has a light sensor and they are constantly sending light values. In addition, there is a fifth device, which works as beacon scanner. The function of this device is to receive all the data coming from the other nodes, decode them, and send them through LoRa using another module that contains that protocol.

Figure 16 describes the devices used for the real scenario according to the wireless technologies mentioned above. A communications sniffer was also used in addition to the application interface www.thethingsnetwork.org. In the middle of the figure, two devices act as coordinating nodes. The distribution of the nodes in real space is shown in Figure 17 where we have an area of approximately 300 m × 300 m that is displayed in Figure 14. We show an approximate radius of coverage of the sensors of 40 m. However, it can vary according to the antenna. Empirically it can be smaller by the amount of collisions according to wireless technology.

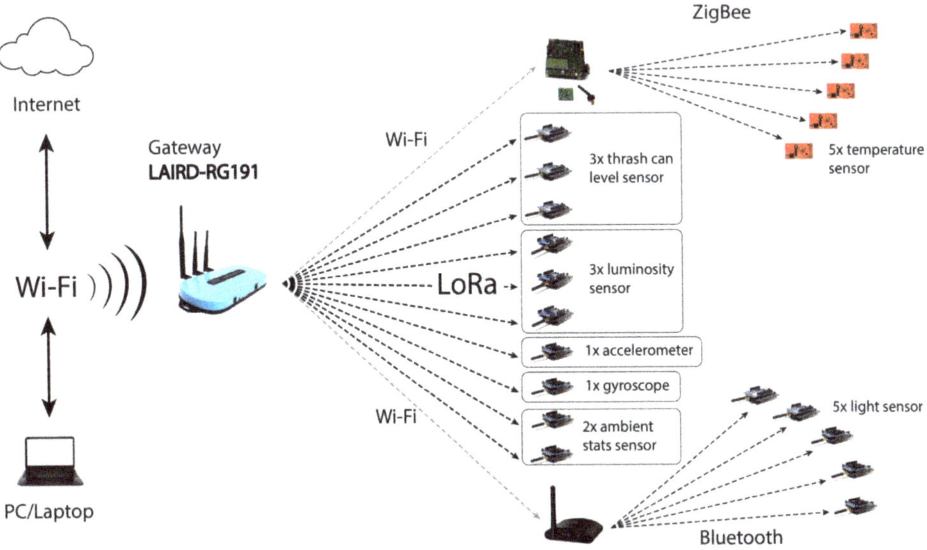

Figure 16. Devices used for the experimental validation [40].

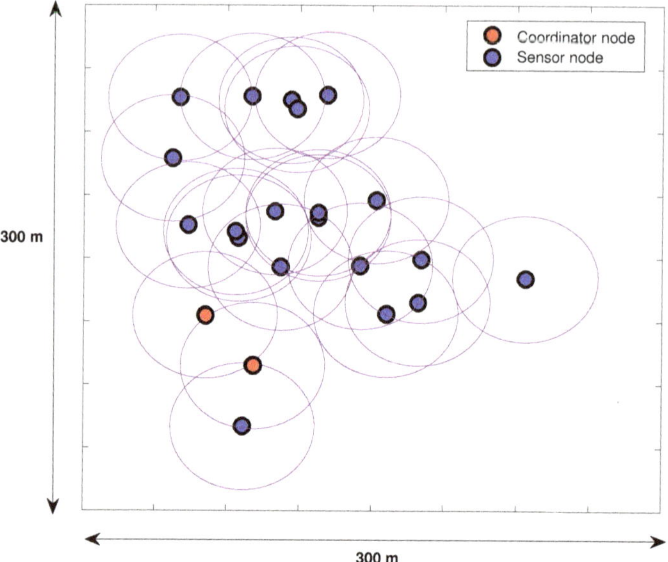

Figure 17. Distribution of nodes in the test space.

Figure 18 shows an example of the plot of the packets with some wireless technology, in this case, LoRa. The package and console display interface is www.thethingsnetwork.org, which allows us to observe the quality parameters of links, frequencies, lost packets, etc. Thanks to this analysis and the energy model, we can sharpen the precision of the model and further refine the types of energy to detect possible inefficient expenses of actions in the network nodes.

```
 1  {
 2      "gw_id": "eui-c0ee40ffff294455",
 3      "payload": "QNAfASaAHAABSf7SJiI=",
 4      "f_cnt": 28,
 5      "lora": {
 6         "spreading_factor": 7,
 7         "bandwidth": 125,
 8         "air_time": 46336000
 9      },
10      "coding_rate": "4/5",
11      "timestamp": "2019-08-13T22:19:27.036Z",
12      "rssi": -43,
13      "snr": 9.25,
14      "dev_addr": "26011FD0",
15      "frequency": 904700000
16  }
```

Figure 18. Example of frame for performance metrics.

Table 6 shows three useful performance metrics for analyzing the behavior of a network. The results for three different types of protocol are described: reagent (AODV), hybrid (MPH), and proactive (ZTR). The overall performance is better in the MPH protocol because it has route redundancy yet, it does not have so many routes to generate too much overhead and network collisions. We also note that the simulator has the ability to accurately reproduce (approximately 2% difference) the real scenario under the specific conditions on each node according to the wireless technology used. In this way, different routing protocol rules suitable for coexisting networks in a wireless medium can be tested and energy optimization models can be generated according to the technology used and the target application. This makes the simulator an effective tool in predicting packet routing and power consumption models.

Table 6. Performance metrics for the real scenario and the simulation.

Metric	Real Scenario			Simulation		
	AODV	MPH	ZTR	AODV	MPH	ZTR
Energy (J)	265.87	140.51	166.64	262.37	142.94	160.73
Delay from the furthest node (s)	1.9456	1.1567	1.0123	1.8955	1.1166	1.0573
PDR (%)	72	90	81	70	88	79

Using this simple energy model, we can implement energy saving techniques in some of the activities carried out by the nodes and we are capable of quantifying their impact on the total expenditure. For Figure 19, we implemented the MPH protocol in order to contrast the energy expenditure of each type of energy in the model under both stable and adverse conditions. For the four different wireless technologies, the rules of the MPH hybrid protocol were used. We observed that when there were adverse conditions such as shutting down, 15% of the nodes were turned off for 10 min every two hours, implying that the energies that have the greatest consumption impact are: CSMA energy, transmission energy, and receiving energy, increasing their value approximately by 40%. In these kind of situations, the model allows the use of routing protocols suitable for the operation of the network.

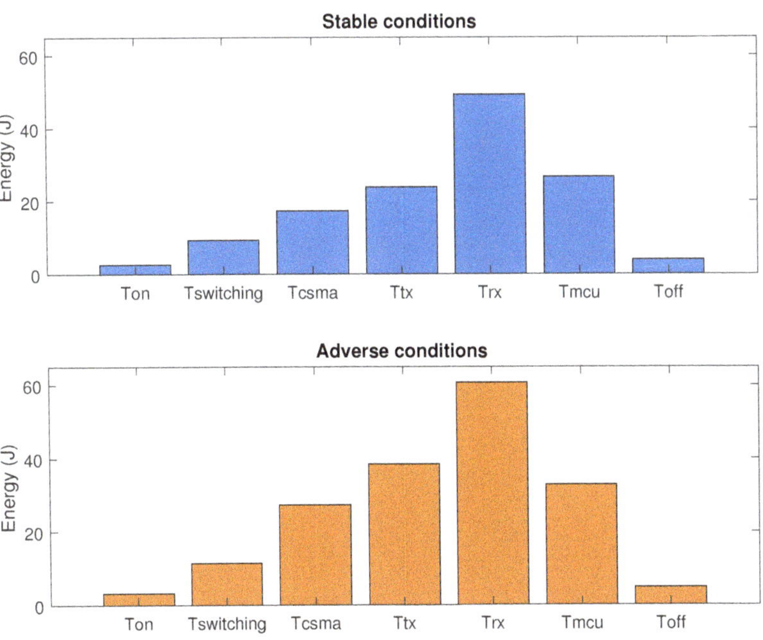

Figure 19. Energy under stable and adverse conditions for 22 nodes for each type of the energy model.

3.2. Extrapolation of the Experimental Scenario

The experiments carried out with 22 sensors were then expanded to 100 sensors and 200 sensors. This is in order to predict the results of the energy model by means of an event-based simulator, programmed in C ++ and previously tested in [10].

Next, it is interesting to analyze how the energy model behaves when the network begins to grow in its number of nodes. Table 7 shows three representative protocols out of the six we have studied in this work. Two predominant types of energies were analyzed based on the scalability of the network, from 22 to 300 nodes. Simulations were run for a full day and for nodes which parameters are detailed in Table 3.

Table 7. Energy model scalability for AODV, PEGASIS, and MPH protocols.

Protocol	22 nodes	100 nodes	150 nodes	200 nodes	250 nodes	300 nodes
AODV $E_{TX} = 186.84J$	$E_{RX} = 151.35J$	$E_{RX} = 301.73J$ $E_{TX} = 404.11J$	$E_{RX} = 403.23J$ $E_{TX} = 572.45J$	$E_{RX} = 567.85J$ $E_{TX} = 681.05J$	$E_{CSMA} = 425.02J$ $E_{TX} = 703.42J$	$E_{CSMA} = 431.86J$ $E_{TX} = 721.33J$
	$E_{TX} = 298.77J$					
PEGASIS $E_{TX} = 86.23J$	$E_{RX} = 99.32J$	$E_{RX} = 125.23J$ $E_{TX} = 110.48J$	$E_{RX} = 200.61J$ $E_{TX} = 144.67J$	$E_{RX} = 301.53J$ $E_{TX} = 211.43J$	$E_{CSMA} = 342.55J$ $E_{TX} = 368.26J$	$E_{CSMA} = 415.55J$ $E_{RX} = 402.56J$
	$E_{TX} = 98.45J$					
MPH $E_{TX} = 90.46J$	$E_{RX} = 73.94J$	$E_{RX} = 101.45J$ $E_{TX} = 193.44J$	$E_{RX} = 192.37J$ $E_{TX} = 203.78J$	$E_{RX} = 265.11J$ $E_{TX} = 287.09J$	$E_{RX} = 325.38J$ $E_{TX} = 382.31J$	$E_{RX} = 404.88J$ $E_{TX} = 440.03J$
	$E_{TX} = 111.52J$					

Results in Table 7 show the two predominant energies for each network and for each of the three protocols studied. Here, we highlight the utility of the proposed model, which breaks down the total energy into the basic energy types of a node. This information becomes relevant when analyzing the skeleton of what is happening in the node with respect to performance metrics such as overhead, collisions, packet loss, interference, complexity of handling neighbor or routing tables, etc.

Analyzing the data, we observe that, when the network is large (about 300 nodes), the AODV and PEGASIS protocols present predominantly CSMA and TX energies. This may suggest that PEGASIS is not a protocol for networks that are too large because the route chain to reach the destination node becomes complex and packets can be lost. Below the 200 nodes, the three protocols under evaluation present the energies of RX and TX as preponderant. Visualizing in detail the network of 100 nodes, the difference between AODV and MPH is 67 % in favor of MPH, and the difference between MPH and PEGASIS is 58% in favor of MPH for the reception energy. This is an example that shows that MPH has a hierarchical topology and that the neighbor tables of the nodes become more manageable thanks to the periodicity of their update. The AODV protocol has the highest energy expenditure and this may be due to the fact that it has a greater number of control packets and if the routes expire or become obsolete, the nodes must start the entire route request process again. In this case, when the networks increase in size, the proposed energy model is very useful because, according to the characteristics of the network, it can establish weak points of behavior and detect possible gaps in energy loss. For example, if CSMA power is increased too much, this may be showing that the communication channel is continuously busy or that this is combined with multiple packet retransmissions, which implies that the processing time increases and the performance of the device is impaired. Then, the breakdown of energies performed by the model can detect abnormal increases or decreases in the behavior of the nodes, as occurs, for example in the PEGASIS protocol when the network is very large.

4. Conclusions

Unlike other types of wireless networks, Wireless Sensor Networks (WSNs) involve low-cost and low-processing devices, which send information to a collector node or base station (coordinator node). Due to the small size of nodes, the saving of energy consumption is vital since it is very difficult to recharge batteries and these networks aim to achieve maximum efficiency in the delivery of information in the harshest environments.

This paper has proposed the design and implementation of a simple and easy to develop energy model, which aim is to observe locally and globally the energy of the nodes in a network under almost any routing protocol. It has been verified that this model yields clear and concrete results for the main tasks performed by a node in the network. Based on this analysis, it could detect some anomalous behavior and know exactly in which phase of execution a problem is happening.

Another fundamental contribution is the analysis of performance metrics, not so common or obvious, because during the transmission of information, the processing of routes, and the observation of the means of communication, the nodes present valuable information when some strange behavior is occurring and this is immediately reflected in metrics such as: retransmissions of packets, listener retries to the communication channel, delays, overload of control packets, hop numbers, valid routes to a destination, among others. These metrics are directly related to energy and to the application of the proposed model. Changes or different behaviors are clearly observed in specific areas of the network.

The MPH protocol works very well in terms of processing, efficient information delivery, and low energy consumption, maintaining route redundancy. The detailed energy model shows that AODV and DSR have route backups and extensive routing tables, therefore a node can reach almost any point in the network. The model also shows that ZTR has low redundancy, is prone to failures, and has a small number of valid routes. However, it is simple, fast, and consumes a small amount of energy. The combination of a hierarchical topology with auto-configuration mechanisms and maintenance of the MPH protocol makes the nodes capable of optimizing network processes, reduce delays by up to 25%, take short routes to the destination, and reduce network overload even in a 30%. All this is reflected in the successful delivery of information. In addition, the proposed model allows us to understand that between MPH and LEACH and PEGASIS there is only a difference of 3% and 2% energy savings for the last two protocols. Thus, the model analyzes the energy impact of each type of energy for optimization of the algorithm in various protocols of the literature [5–10]

Author Contributions: C.D.-V.-S. developed the energy algorithm, built the simulator, prepared and executed the simulations, interpreted and analyzed the results, designed the methodology and drafted the manuscript. C.M.-P. supervised the research methodology and the approach of this work, he performed the formal analysis. J.A.N.-F. reviewed, interpreted and drafted the simulation results, he also strongly contributed to the design of the energy scheme. R.V. was involved in the analysis of the energy model under routing protocols and he run validation, he reviewed the methodology and the manuscript. A.R.-S. worked in the formal analysis and the manuscript. All authors have read and agreed to the published version of the manuscript.

Funding: This research received no external funding.

Conflicts of Interest: The authors declare no conflict of interest.

References

1. Kurt, S.; Yildiz, H.U.; Yigit, M.; Tavli, B.; Gungor, V.C. Packet size optimization in wireless sensor networks for smart grid applications. *IEEE Trans. Ind. Electron.* **2017**, *64*, 2392–2401. [CrossRef]
2. Han, G.; Liu, L.; Jiang, J.; Shu, L.; Hancke, G. Analysis of energy-efficient connected target coverage algorithms for industrial wireless sensor networks. *IEEE Trans. Ind. Inform.* **2017**, *13*, 135–143. [CrossRef]
3. Kabila, A.; Murugan, A. Efficient Energy Performance of the Wireless Sensor Networks and Cross Layer Optimization. *Asian J. Appl. Sci. Technol. (AJAST)* **2017**, *1*, 55–58.
4. Del-Valle-Soto, C.; Mex-Perera, C.; Monroy, R.; Nolazco-Flores, J.A. On the routing protocol influence on the resilience of wireless sensor networks to jamming attacks. *Sensors* **2015**, *15*, 7619–7649. [CrossRef] [PubMed]
5. Perkins, C.; Belding-Royer, E.; Das, S. *Ad Hoc On-Demand Distance Vector (AODV) Routing*; Technical Report; IETF: Fremont, CA, USA, 2003.
6. Maltz, D.A.; Broch, J.; Jetcheva, J.; Johnson, D.B. The effects of on-demand behavior in routing protocols for multihop wireless ad hoc networks. *IEEE J. Sel. Areas Commun.* **1999**, *17*, 1439–1453. [CrossRef]
7. Zigbee, A. Zigbee specification. In *ZigBee Document 053474r13*; ZgBee Standards Organization: San Ramon, CA, USA, 2006.
8. Bhat, G.; Sreenivasan, A. Review on energy optimization and cluster based routing protocol in WSN. *Int. Res. J. Eng. Technol. (IRJET)* **2019**, *6*, 101–103.

9. Trigunait, C.K.; Prabha, S. A Novel Energy Efficient Security Protocol In WSN. *Int. J. Inf. Technol. (IJIT)* **2019**, *5*, 1–4.

10. Del-Valle-Soto, C.; Mex-Perera, C.; Orozco-Lugo, A.; Galvan-Tejada, G.M.; Olmedo, O.; Lara, M. An efficient multi-parent hierarchical routing protocol for WSNs. In Proceedings of the 2014 IEEE Wireless Telecommunications Symposium (WTS), Washington, DC, USA, 9–11 April 2014; pp. 1–8.

11. Zaman, N.; Tang Jung, L.; Yasin, M.M. Enhancing energy efficiency of wireless sensor network through the design of energy efficient routing protocol. *J. Sens.* **2016**, *2016*, 9278701. [CrossRef]

12. Ghaffari, A. An energy efficient routing protocol for wireless sensor networks using A-star algorithm. *J. Appl. Res. Technol.* **2014**, *12*, 815–822. [CrossRef]

13. Ordóñez, F.; Krishnamachari, B. Optimal information extraction in energy-limited wireless sensor networks. *IEEE J. Sel. Areas Commun.* **2004**, *22*, 1121–1129. [CrossRef]

14. Zhou, H.Y.; Luo, D.Y.; Gao, Y.; Zuo, D.C. Modeling of node energy consumption for wireless sensor networks. *Wirel. Sens. Netw.* **2011**, *3*, 18. [CrossRef]

15. Gu, Y.; Ren, F.; Ji, Y.; Li, J. The evolution of sink mobility management in wireless sensor networks: A survey. *IEEE Commun. Surv. Tutor.* **2016**, *18*, 507–524. [CrossRef]

16. Ren, J.; Zhang, Y.; Zhang, K.; Liu, A.; Chen, J.; Shen, X.S. Lifetime and energy hole evolution analysis in data-gathering wireless sensor networks. *IEEE Trans. Ind. Inform.* **2016**, *12*, 788–800. [CrossRef]

17. Ephremides, A. Energy concerns in wireless networks. *IEEE Wirel. Commun.* **2002**, *9*, 48–59. [CrossRef]

18. Li, J.; Mohapatra, P. An analytical model for the energy hole problem in many-to-one sensor networks. In Proceedings of the IEEE Vehicular Technology Conference, Dallas, TX, USA, 28 September 2005; Volume 62, p. 2721.

19. Bouaziz, M.; Rachedi, A. A survey on mobility management protocols in Wireless Sensor Networks based on 6LoWPAN technology. *Comput. Commun.* **2016**, *74*, 3–15. [CrossRef]

20. Sergiou, C.; Vassiliou, V.; Paphitis, A. Hierarchical Tree Alternative Path (HTAP) algorithm for congestion control in wireless sensor networks. *Ad Hoc Netw.* **2013**, *11*, 257–272. [CrossRef]

21. Nezhad, A.A.; Miri, A.; Makrakis, D. Location privacy and anonymity preserving routing for wireless sensor networks. *Comput. Netw.* **2008**, *52*, 3433–3452. [CrossRef]

22. Wang, D.; Mukherjee, M.; Shu, L.; Chen, Y.; Hancke, G. Sleep scheduling for critical nodes in group-based industrial wireless sensor networks. In Proceedings of the 2017 IEEE International Conference on Communications Workshops (ICC Workshops), Paris, France, 21–25 May 2017; pp. 694–698.

23. Loo, J.; Mauri, J.L.; Ortiz, J.H. *Mobile ad hoc Networks: Current Status and Future Trends*; CRC Press: Boca Raton, FL, USA, 2016.

24. Reddy, D. A Review-Efficiency of Energy Clustering and Routing in Wireless Sensor Networks. *Int. J. Adv. Technol.* **2019**, *10*, 2.

25. Khoury, R.; Dawborn, T.; Gafurov, B.; Pink, G.; Tse, E.; Tse, Q.; Almi'Ani, K.; Gaber, M.; Röhm, U.; Scholz, B. Corona: Energy-efficient multi-query processing in wireless sensor networks. In Proceedings of the International Conference on Database Systems for Advanced Applications, Tsukuba, Japan, 1–4 April 2010; pp. 416–419.

26. Bennani, M.T.; Zbakh, A. Leach routing protocol for image transfer using Castalia simulator. In Proceedings of the TELECOM'2019 & 11 emes JFMMA, Saidia, Morocco, June 2019.

27. Heinzelman, W.B.; Chandrakasan, A.P.; Balakrishnan, H. An application-specific protocol architecture for wireless microsensor networks. *IEEE Trans. Wirel. Commun.* **2002**, *1*, 660–670. [CrossRef]

28. Sneha, K.; Kamath, R.; Balachandra, M.; Prabhu, S. New Gossiping Protocol for Routing Data in Sensor Networks for Precision Agriculture. In *Soft Computing and Signal Processing*; Springer: Singapore, 2019; pp. 139–152.

29. Toor, A.S.; Jain, A. Energy Aware Cluster Based Multi-hop Energy Efficient Routing Protocol using Multiple Mobile Nodes (MEACBM) in Wireless Sensor Networks. *AEU-Int. J. Electron. Commun.* **2019**, *102*, 41–53. [CrossRef]

30. Singh, S.; Raghavendra, C.S. PAMAS—Power aware multi-access protocol with signalling for ad hoc networks. *ACM SIGCOMM Comput. Commun. Rev.* **1998**, *28*, 5–26. [CrossRef]

31. Ekal, H.H.; Abdullah, J.B. Energy Provisioning Technique to Balance Energy Depletion and Maximize the Lifetime of Wireless Sensor Networks. *Energy* **2016**, *7*. pp. 276–282.

32. Chen, Z.; Liu, A.; Li, Z.; Choi, Y.j.; Li, J. Distributed duty cycle control for delay improvement in wireless sensor networks. *Peer-to-Peer Netw. Appl.* **2017**, *10*, 559–578. [CrossRef]

33. Tsuchie, K.; Yao, T.; Kubo, Y.; Nakamura, N.; Nozaki, M. Parent node discovery time reduction algorithm for low-power wireless sensor networks. In Proceedings of the 2017 Ninth International Conference on Ubiquitous and Future Networks (ICUFN), Milan, Italy, 4–7 July 2017; pp. 176–181. 10.1109/ICUFN.2017.7993770.

34. *Wireless LAN Medium Access Control (MAC) and Physical Layer (PHY) Specifications*; IEEE P802. 11 D3; IEEE Standards Dept.: New York, NY, USA, 1996.

35. *A True System-on-Chip Solution for 2.4-GHz IEEE 802.15. 4 and ZigBee Applications*; CC2530 Datasheet; Texas Instruments: Dallas, TX, USA, 2011.

36. Bouguera, T.; Diouris, J.F.; Chaillout, J.J.; Jaouadi, R.; Andrieux, G. Energy consumption model for sensor nodes based on LoRa and LoRaWAN. *Sensors* **2018**, *18*, 2104. [CrossRef] [PubMed]

37. Kim, C. *Measuring Power Consumption of CC2530 with Z-Stack*; Application Note AN079; Texas Instruments: Dallas, TX, USA, 2012.

38. Jurdak, R.; Ruzzelli, A.G.; O'Hare, G.M. Radio sleep mode optimization in wireless sensor networks. *IEEE Trans. Mob. Comput.* **2010**, *9*, 955–968. [CrossRef]

39. Shu, L.; Mukherjee, M.; Wang, D.; Fang, W.; Chen, Y. Prolonging global connectivity in group-based industrial wireless sensor networks. In Proceedings of the 16th ACM/IEEE International Conference on Information Processing in Sensor Networks, Pittsburgh, PA, USA, 18–21 April 2017.

40. Del-Valle-Soto, C.; Valdivia, L.; Velazquez, R.; Rizo-Dominguez, L.; Lopez-Pimentel, J.C. Smart Campus: An Experimental Performance Comparison of Collaborative and Cooperative Schemes for Wireless Sensor Network. *Energies* **2019**, *12*, 3135. [CrossRef]

Article

Smart Campus: An Experimental Performance Comparison of Collaborative and Cooperative Schemes for Wireless Sensor Network

Carolina Del-Valle-Soto [1,†], Leonardo J. Valdivia [1,*,†], Ramiro Velázquez [2], Luis Rizo-Dominguez [3,†] and Juan-Carlos López-Pimentel [1]

[1]   Facultad de Ingeniería, Universidad Panamericana, Álvaro del Portillo 49, Zapopan, Jalisco 45010, Mexico
[2]   Facultad de Ingeniería, Universidad Panamericana, Josemaría Escrivá de Balaguer 101, Aguascalientes 20290, Mexico
[3]   ITESO, Universidad Jesuita en Guadalajara, San Pedro Tlaquepaque 45604, Mexico
*   Correspondence: lvaldivia@up.edu.mx; Tel.: +52-33-13682200
†   These authors contributed equally to this work.

Received: 20 July 2019; Accepted: 11 August 2019; Published: 15 August 2019

Abstract: Presently, the Internet of Things (IoT) concept involves a scattered collection of different multipurpose sensor networks that capture information, which is further processed and used in applications such as smart cities. These networks can send large amounts of information in a fairly efficient but insecure wireless environment. Energy consumption is a key aspect of sensor networks since most of the time, they are battery powered and placed in not easily accessible locations. Therefore, and regardless of the final application, wireless sensor networks require a careful energy consumption analysis that allows selection of the best operating protocol and energy optimization scheme. In this paper, a set of performance metrics is defined to objectively compare different kinds of protocols. Four of the most popular IoT protocols are selected: Zigbee, LoRa, Bluethooth, and WiFi. To test and compare their performance, multiple sensors are placed at different points of a university campus to create a network that can accurately simulate a smart city. Finally, the network is analyzed in detail using two different schemes: collaborative and cooperative.

Keywords: energy consumption; wireless sensor network; cooperation; collaboration

---

1. Introduction

The Internet of Things (IoT) is based on the connection of multiple devices to the Internet and data sharing between them. Data is first sent to the cloud where it is processed using analytics, and then sent back to other devices. This process depends completely on a centralized architecture. A Wireless Sensor Network (WSN) is considered the most critical element of the IoT model. In the context of IoT, these wireless networks play an essential role in increasing the ubiquity of networks [1] since wireless technology is the fundamental way in which "intelligent objects" communicate with each other and the Internet. In this sense, WSNs allow IoT scalability and provide enough functionality to support its integration with the current Internet architecture. Moreover, it is essential to study the scalability and the adaptation methods of the network in the face of packet transmission failures and topology changes [2].

IoT provides smarter services by the interconnection of various objects. Providing intelligent services requires data to be collected from different places, areas, and devices. Hence, in these kinds of applications, energy consumption is a key factor; sensors can be in remote zones and they can be challenging to access, so it is not possible to replace their batteries continuously. Due to the limitations of battery life, the nodes are designed to save as much energy as possible and most of the time they are in sleep mode (low power consumption mode). It is thus essential to know the factors that have an impact on battery life [3]. Regarding the wireless medium, aspects that contribute to the deterioration of the information transmission must be taken into account as the channel parameters depend directly on the characteristics of the transmission medium. These characteristics affect the medium and the duration of the useful life of the batteries because there might be connections and disconnections of some nodes in the network. The conditions of the medium also influence the packet retransmission and listen attempts to the channel. This causes the routing protocol to increase its control messages and deteriorate its overhead [4].

WSNs are multi-functional, low-cost, and low-power networks that rely on communications among nodes or from sensor nodes to one or more sink nodes. Sink nodes, sometimes called coordinator nodes or root nodes, may be more robust and have larger processing capacity than other kinds of nodes. Sensor networks can be widely used in various environments even the hostile ones. Some of the many applications of WSNs are in the medical field, agriculture, monitoring, and detection, automation, and data mining. The key element of the IoT is that different devices are connected within the same environment being able to transmit and receive information about their immediate environment and interaction with users. The accuracy of the location allows interactions and data sharing between devices that automate and simplify tasks that generate comfort and facilitate daily work practices. Continuing developing technologies that combine IoT and location will lead to the development of better services for automation and support of daily life tasks.

Adaptive methods for wireless networks have been studied because of the introduction of many technologies for using different technical standards [5]. The concepts of coexistence and cooperation are currently booming in the literature. Both have to do with how the nodes relate to each other and to the network itself. A balanced network seeks to match the load of information traffic and routes, adapting the rules of routing protocol by means of control packages. Coexistence and cooperation offer an idea on how the device puts its functions at the service of the network regardless of the technologies or protocols in use. One of the challenges when analyzing these concepts is interference. There are different metrics that indicate the quality of links in wireless networks.

To analyze the terms that describe the relationship of the devices among them and with the network, it is important to define the degree of intelligence that these nodes possess, a term that is included in technological convergence. When talking about heterogeneous environments, we can mention different network topologies so that the devices exist in an environment that has different characteristics.

- Cooperative networks use self-configuration capabilities to dynamically adapt on demand, since they can respond to the needs of a specific user, within the policies defined by the operator, while optimizing the general resources of network.
- Cognitive networks work on the physical level of the protocol stack, handling emission frequencies and modulation parameters. Cognitive networks exhibit relevant characteristics such as: (1) identification of spectrum access opportunities, (2) selection of frequency bands to be used, (3) coordination among users for access to spectrum, and (4) spectral mobility.

The main disadvantage of cooperative networks lies in the shared management and interdependence of networks that are in an environment with common conditions. Cognitive networks have an agile cognitive process with which they can determine the current conditions of the network, in order to plan, decide, and act. The network can learn from these adaptations to make future decisions taking into account the final objectives: to develop communication protocols.

Motivation

The motivation of this work is to evaluate the performance of different wireless technologies, such as: Zigbee, LoRa, Bluetooth Low Energy, and WiFi. These technologies have been widely spread in industrial applications due to their low cost and low power consumption. The comparative relations of performance parameters intrinsic to different routing protocols are established to achieve the interconnection of networks of sensing devices with communication Long Range Wide-area network (LoRaWAN, sub-1 GHz) and BLE, in order to emulate an intelligent campus. Each network will have its own sniffer for the acquisition and analysis of the packets sent between the nodes, connectivity between BLE and LoRa devices through a gateway with the possibility of sending data to the cloud.

2. Related Work

The current revolution towards a universal connection between things, named IoT, is considered part of the Internet of the future. The long-term plan is to have approximately 28 billion connected devices by 2021 [6].

IoT is currently considered to be one of the most outstanding areas of future technology. In the following subsections we explain about the global applications in this area from smart houses, smart campuses, and smart cities; then we explain some research work focused on identifying the main challenges dealing with IoT. We conclude the section detailing previous works that have performed comparative studies between wireless protocols.

2.1. IoT Applications

A typical IoT application is a smart house where the refrigerator can report the current status of the food; the temperature of each room sets according to the preferences of the people occupying such space; and all devices working on identifying possible catastrophes such as a gas leak.

Another application is a smart campus which can use the new generation of information technology such as IoT, cloud computing, and big data to perceive, store, manage, and analyze all the key information of campus system; all of these can serve, for example, to save energy and assisting the staff such as faculty, students and administrative in decision making [7].

The evolution of IoT, starting from smart house, through smart campus and joining with other technologies such as big data, data analysis, among others, have paved the way to smart cities. This concept brings a bigger picture from the houses into a whole city where public transportation, energy, water supply, and environmental factors are at stake.

For example, there is a large number of different devices designed for IoT, but also modern smartphones can work since they include multiple sensors, communication standards, and can store data. The authors in [8], have looked at a different perspective by using smartphones to create the next generation of civil infrastructure monitoring systems. However, Esposte et al. [9], have identified that there are not enough tools to design the cities of the future. Therefore, they created InterSCity, which is a smart city platform that focuses on collaborative development of services, applications, and systems. Unfortunately, it is complex and requires a lot of time and a large budget to do enough testing to actually identify opportunities.

Consequently, researchers are now focusing on working on smart campuses, which is a bridgework between smart homes and cities.

A campus is large enough to provide challenges related to device positioning and at the same time it allows a safe environment to collect data from the parking lot, the usage of rooms, the amount of water and electrical energy during the day and monitor other aspects as well.

Thus, the authors in [10] used an agent-based computing paradigm to gather data from their smart university campus where they could test communication issues and solve them as a decentralized system.

Another example is [7] where key data is stored and analyzed to provide useful information to teachers, students, and staff. Monitoring a classroom to identify how much time is used and by how many people, can help take better decisions on how to assign classrooms to different groups in such a way that it is optimized.

Lghoul et al. [11] have mainly focused on profiling energy usage in campus buildings, setting power distribution system architecture, and pinpointing key micro-grid components. They propose a general Microgrid (MG) testbed and simulate the operation of proposed MG model/architecture. They delineate relevant pros and cons towards a futuristic real-world/physical MG deployment in a university campus.

IoT is gaining a special moment and the application in smart campuses (respectively smart cities) requires essential technologies for the deployment of successful products and services as described by Lee and Kyoochun [12]. Such essential technologies are:

(a)   Radio frequency identification (RFID);
(b)   Middleware;
(c)   IoT software application;
(d)   Cloud computing; and
(e)   Wireless sensor networks (WSN).

In the following subsections we put our attention in WSN.

2.2. Energy and Power Consumption

There are challenges to solve in IoT, for instance, the amount of energy that each accessory requires and the coexistence of thousands of wireless devices that may use different technologies for communicating [13]. In particular, coexistence is a severe threat to IoT, since the devices that are trying to communicate may collide with each other resulting in data lost, rendering the sensors useless. While there is substantial work on coexistence for Bluetooth, WiFi, and IEEE 802.15.14 standards [14], the forecast includes thousands of devices in a narrow range. This also considers the application of IoT to manufacturing and much more, which is called Industrial Internet of Things (IIoT), and it is based on the possibility of creating large-scale deployments. Consequently, some approaches include using a modified Medium Access Control (MAC) protocol. Nevertheless, there are multiple limitations as described by [15].

Another challenge of IoT is power consumption, in this sense, Mahmoud and Mohamad [16], have presented a study of wireless technologies for IoT applications. Their study was focused on the importance of using low-power wireless techniques by introducing a comparative between different low-power wireless communication techniques such as ZigBee, Low-Power WiFi, 6LowPAN, and LPWA.

2.3. Signal Interference and Collision

Interference and collisions are also important challenges in IoT. An approach sensing the spectrum to identify interference in IIoT devices is [17], but in practical use it takes a large amount of time to identify the issue and provide a solution. A more recent method [18] uses support vector machines that can sense under 300 ms and classifies external interference. In other words, a management system is required to minimize the amount of collisions and increase the effectiveness, such as [19] that uses a self-learning system based on reinforcement learning. Adeyemi et al. presented in [20] a robust data exploration study performed on daily Internet data traffic generated in a smart university campus for 12 consecutive months.

2.4. Comparison between Wireless Protocols

The comparison of different wireless technologies to decide which one is the best, has been very attractive for a sector of researchers. For example, the authors in [21], provide a comparative study of Bluetooth, Ultra-wideband (UWB), ZigBee, and WiFi wireless communication standards. They presented an overview evaluating the main features and behaviors of the standards in terms of various metrics: transmission time, data coding efficiency, complexity, and power consumption. The work did not draw any conclusion regarding which one is better because the authors concluded that more factors must be considered such as network reliability, roaming capability, recovery mechanism, chipset price, and installation cost.

Other work focused on finding a solution to the problem related to the selection criteria of a better wireless communication technology face up to the constraints imposed by the intended application and the evaluation of its key features is that of Chakkor Saad et al. [22]. They presented a comparative performance analysis of the following wireless protocols: WiFi, WiMax, UWB, Bluetooth, ZigBee, ZigBeeIP, and GSM/GPRS. They developed a quantitative evaluation with respect to transmission time, the data coding efficiency, the bite error rate, and the power and energy consumption. Their conclusion is that in order to determine which one is the most suitable, other factors must be taken into account such as network reliability, the link capacity between several networks with different protocols, security, the chipset price, the conformity with the application and the cost of installation.

More recently, Naidu and Kumar [23] have carried out a description on the wireless technologies importance, features, and a comparison about Bluetooth, ZigBee, WiFi, and Z-Wave; mainly focused on Self-organizing/Optimization Networks (SONs). The work of these authors was focused on home automation devices, integrating them with a smart hub. The authors describe WiFi SON, they concluded that it is a guaranteed network, offering quality of service and eliminating human interventions.

One of the characteristics of this type of WSN/IoT networks is the sleeping techniques that can be applied to optimize the energy consumption of the sensors. These types of techniques help to reduce the power of each part of the node, a task that consists essentially of turning off or bringing the device to a low-power mode when it is not used, while when in use, it is activated or awake. By reducing the consumption of each part of the node, the overall consumption is reduced and, therefore, the battery life is extended. In this work, sleep techniques are not properly applied; however, these mechanisms can resemble the connections and disconnections of the nodes that we analyze as a performance metric. Similarly, it must be taken into account that when a node falls asleep it is not necessarily disconnected from the network, but in some cases, depending on the applications, the nodes can temporarily turn off their microcontroller unit and thus, optimize their energy to the maximum and the power of the network. A recent approach to reduce the energy consumption in WSNs is through the setting of sleep scheduling. In addition, a relevant proposal is minimizing the number of nodes to cover a constrained area. Good results in terms of complexity, working-node ratio, scalability, and the time of network duration were obtained in [24].

Unlike the previously mentioned works, our research is concentrated in offering a comparison under a collaborative and cooperative scheme between the following wireless technologies: Zigbee, LoRa, Bluetooth Low Energy, and WiFi. These technologies have gained wide acceptation in industrial applications due to their low cost and low power consumption. Our experimental scenery is a campus, since we are looking for emulating a smart campus.

3. Wireless Communication Networks

We define a wireless communication network (WCN) as the association of wireless elements (WE) to share information with a specific task. The nature of WE defines the association schemes and the network features. A WE can be defined as an electronic artifact that possess sensing, acting, signal processing, and/or communication abilities to perform a specific work [25].

Technology integration in WE allows exploitation of different platforms for establishing WCNs. The network features are defined by the nature, status (Active/No Active), capacities, and relationship of WE considering their physical and technological limitations such as size, power transmission, energy consumption, computing capacity, frequency of operation, sensibility of interference, etc.

The networks are based on schemes, models, and protocols that establish the technology, transmission agreements, architecture, and management. This way, we can describe the association schemes according their functions. Two classes of schemes can be distinguished.

- *Collaborative scheme.* It consists of sharing resources and functions only if the WE has availability and does not jeopardize its operation, i.e., WE priorities are over Wireless Network (WN) priorities.

There are no strong commitments among their elements [26].

- *Cooperative scheme.* Its scheme is defined to share assigned resources and perform specific functions for all the elements in the networks. All the tasks must be performed through strict rules or protocols, i.e., WN priorities are over WE priorities. In this case there are strong commitments among their members [27].

WN can be classified as a WCN considering different aspects such as nature, management strategy, topology, coexistence scheme, and type of WE. Figure 1 shows a chart with basic concepts and aspects about WCN.

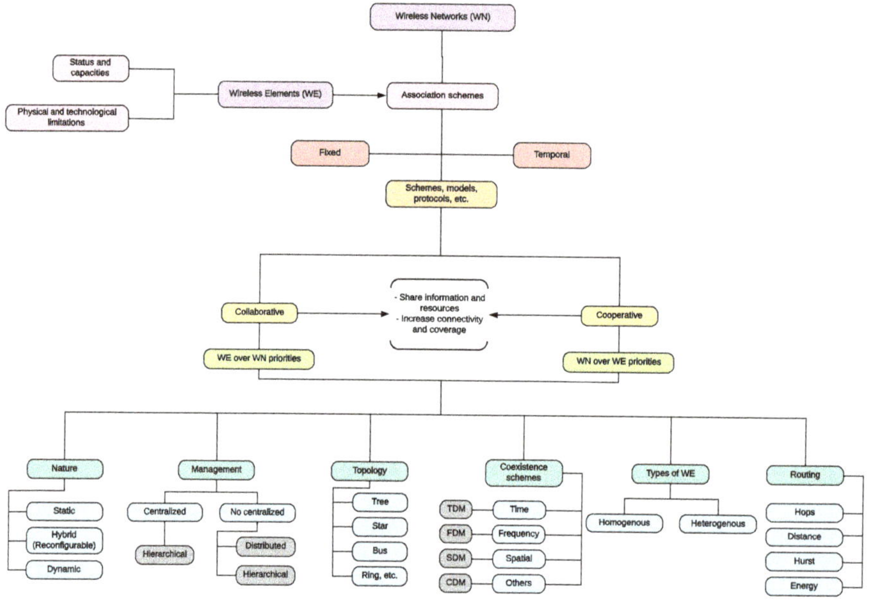

Figure 1. Concepts and aspects of Wireless Communication Networks.

The design of a WCN is motivated or influenced by some of the following requirements: Redundancy of information: the high density of devices makes the data obtained in one of them redundant with respect to other nodes of its environment. Limitation of resources: design and implementation of wireless networks must take into account resource limitations: energy, power, memory, and bandwidth. In addition to being generally limited in size, the devices will depend on their batteries and the power they can extract from the environment for their operation. Topology and dynamic environments. The conditions in which a sensor network is deployed is not fixed, but there may be node movement and even Disappearance or addiction of others. Wireless networks must be able to reconfigure themselves autonomously. Unreliable transmission medium: The use of wireless communications has a considerably higher error rate than the wired communications. Security and privacy: these factors are especially important in military and surveillance applications. Therefore, denial of service, intrusion, or data manipulation attacks on these applications must be foreseen.

Networking is the mechanism to construct networks, where each node obtains information about their neighbors and the rules of communication that define the schemes, strategies, and protocols used to establish communication model among their members. Thus, we can describe, in wide sense, the networking process in following phases:

- *Observation.* Nodes must explore the channel conditions with the purpose of perceiving the existence of other nodes and noise. For this reason, each node must continuously sense the channel.
- *Recognition and identification.* Nodes can recognize information signals from the channel noise, as well as being a process that allows identification of the neighbors and the rest of members of networks.
- *Association scheme.* At this phase, the agreements about the guidelines and rules define the relationship of each node with the rest of nodes.
- *Transmission agreements.* Transmission power, modulation format, and frequency of operation are some examples of parameters agreed in this phase.
- *Topology and routing.* The interconnection network defines the topology, i.e., the available communication links among nodes establishing the architecture, while routing defines the paths that follow the information to reach its destination.
- *Management scheme.* There are schemes that ensure the operation, access to channel, and control of transmission parameters.

Presently, dynamic and intermittent behavior of WE represent relevant challenges for networking, so the new requirements of WCN such as flexibility in architecture, mobility, and high traffic play a very important role for modern networks. Reconfigurable ability is considered a prominent network paradigm for future WCN [28]. Besides, high density of WE also represents additional challenges due to closeness of WE and the phenomena that it involves, e.g., multiuser channel access problem, multiuser, and multiple access interferences.

Reconfigurable wireless networks (RWN) are dynamic networks suitable for the frequent changes in topology, connectivity, routing, WE status, and wireless channel conditions. These networks have the ability of modifying their configuration, with or without infrastructure, that establish the allocation decisions; self-organization networks imply an *intelligence* process, which can have centralized or no centralized management, that make them attractive to networks as sensors and ad-hoc.

The access to resources and multiple transmissions in WCN face important challenges such as:

- *Coexistence.* It consists of the study of the presence of WE in a given area, the communication among WE, and their influence in the wireless communication channel. Coexistence schemes ensure the communication of multiple WE through the management and distribution of resources.
- *Energy efficiency.* It refers to the efficient use of energy by WE. Strategies seeking to extend the lifespan of battery due to the limited source of energy. Thus, there are strategies based on efficient routing, schedule, low processing systems, and topology and traffic control which reduce the energy consumption from different aspects.
- *Interference.* The received signals from wireless channel can be distorted due to the presence of interference which can be caused by natural or transmission factors. Hence, techniques for mitigation, avoidance, cancelation, and management interference play an important role.
- *Reception algorithms.* It refers to strategies used to extract the information from the channel. The study of algorithms based on reception and detection strategies are proposed which can be classified according to the techniques and methodologies used [29].
- *Security.* Protocols and strategies that provide privacy and authentication of information to avoid vulnerabilities due to malicious eavesdropping.
- *Complexity.* It refers to the difficulty of resolving computational problems through efficient algorithms.

Communication systems used for RWN use orthogonal or semi-orthogonal multiple access channel techniques to avoid multiple users channel problems. They consider temporal, spectral, and spatial dimensions to separate users. However, their implementation requires to control the access to resources using elaborated schemes and protocols to ensure the communication.

These devices have limited resources such as low power, computing capacity, narrow bandwidth transmission, so on, high sensibility to interference, and autonomous behavior. New trends incorporate a communication system that allows the formation of temporal wireless reconfigurable networks (WRN), i.e., sensors and ad hoc networks, that share information by establishing communication links through technologies and schemes that ensure their operation in free bands or unlicensed, e.g., BLE, ZigBee, and WiFi in the bands 2.4 and 5 GHz.

The high density of WDs leads the multiuser channel problem, which occurs when multiple users share spectrum resources. The multiuser channel is divided in two types, the first is the broadcast channel where a single WD transmits signals to many WD receivers e.g., the relationship between the base station (BS) and user equipment (UEs) in cellular networks. Secondary, the multiple access channel uses schemes that distribute the resources and permit the transmission among many transmitters with one receiver [30,31].

The access to the multiuser channel employs schemes to allocate users with conditions to communicate. The schemes based on random allocation is known as random multiple access (RMA) scheme that is widely used e.g., in satellite networks and large wireless networks without signaling overheads [32]. On the other hand, the schemes based on multiple access allocation assign dedicate channels to users using orthogonal or semi-orthogonal division of resources [31,33]. This way, some techniques based on multiple access schemes have been proposed to provide uniformly the resources considering the temporal, frequency and spatial parameters or combinations to separate users [34,35].

The presence of multiple WDs in a given area is widely studied because it can change the transmission considerations in the wireless channel. Multiple access schemes are useful techniques that require previous agreements. This way, the schemes can be described in temporal, spectral, spatial or combinations.

- *Temporal.* This scheme establishes transmission periods which can be used orderly or opportunistically. For example, time division multiplexing (TDM) techniques are orderly schemes due to the fact that time is divided in intervals with the same length that are assigned to each active user. Cognitive Radio technologies is an opportunistically scheme due to the fact that the channel is used by secondary users only when it is possible without disturbing the primary user transmission.
- *Spectral.* This scheme consists of assigning sub-bands to each communication link. They can be assigned one-per-user or flock-per-user during a communication session, i.e., single-carrier (SC) or multi-carrier (MC) basis systems. Frequency division multiplexing (FDM) and multi-carrier technologies are some examples of this kind of scheme.
- *Spacial.* This scheme consists of spatially separating the resources considering their coverage area forming clusters, cells, or sectors coverage to allow their reuse. These schemes are widely used in satellite communications and wireless mobile telephony systems.

3.1. Energy Efficiency

The efficient use of energy is a very important challenge in WCNs due to the limited source (battery). For this reason, energy considerations have been focused on maximizing the lifetime of the batteries of the WDs that form the WCN. According to [36], the energy efficiency protocols proposed can be classified into four strategies considering the battery outage, the time to unavailability of applications functionality and the time of the first network partitioning, This way, the strategies to energy efficiency are classified as:

1. Energy efficiency routing that reduces energy consumption for each end-to-end transmission.
2. Scheduling the WDs status to save energy and ensure the network and application functionalities.
3. Topology control by turning node transmission power to find the optimum transmission power with the minimum energy consumption that ensures the connectivity.
4. Strategies that reduce the volume of information transferred.

3.2. Interference

The received signals from wireless channel can be distorted due to the presence of interference, which are caused by natural or transmission factors. Hence, techniques for mitigation, avoidance, cancelation, and management of interference play an important role. The interference caused by multiple transmissions are of interest as example.

- Multiple access interference (MAI) occurs when multiple communications links access to channel.
- Multipath interference (MPI) is produced by different propagation paths to reach a specific receiver.
- Multiuser interference (MUI) occurs when simultaneous communication links exist in the same channel from different WDs. Two particular cases can be identified: *Exposed* and *Hidden* terminals.

3.3. Synchronization

It is the process where the WDs define the alignment of transmission parameters, which allows the establishment of communication links, i.e., to define the information necessary to ensure the hookup and linkup among WDs. This process plays an important role given that it has impact in both physical and higher levels.

3.4. Security

The dynamic and flexibility of RWN represent a challenge for implementation of security schemes due to frequent changes in the network status, e.g., sporadic connectivity, variable number of active WDs, temporal topology, etc. In addition, it is desirable to provide authentication and privacy requirements to avoid vulnerabilities due to malicious eavesdropping devices. Commonly, the schemes require associative mechanisms, security protocols which can be proactive or reactive and involve cross layer functions or coding signals [37,38].

3.5. Complexity

It refers to the difficulty to resolve computational problems analyzing the algorithms implemented in the WDs and the use of resources as memory, storage, energy, time, etc., [39]. The challenge in complexity is to implement efficient algorithms for wireless systems that permit to reduce the power consumption and the size of devices while decreasing its manufacturing cost. However, the simplicity must ensure the *quality of service (QoS)* and privacy.

4. Performance Metrics Study

Metrics of the network layer are fundamental because they show the performance and usefulness of a routing protocol. Each routing protocol is designed for specific applications and certain scenarios. These metrics indicate how the bandwidth use is affected by the overhead of the routing protocol in use. In addition, the availability of effective routes and the ability of the network for self-configuration show the capacity of the protocol to recover from topology changes [40]. Recovery times have an impact on the latency in the network, and even though the networks conform with different technologies, it is highly essential to understand and evaluate the performance metrics as shown in [41]. This analysis allows network customization to improve its different aspects and provide better communication.

Therefore, using these metrics the proactive and reactive protocols can be compared. To perform such comparison a simulated and a real scenario were implemented. The next section describes the simulation.

Figure 2 describes performance metrics in a general way. The measurement and control of these performance parameters provide the network optimization because some parameters influence others, allowing the information in a wireless network to be delivered reliably.

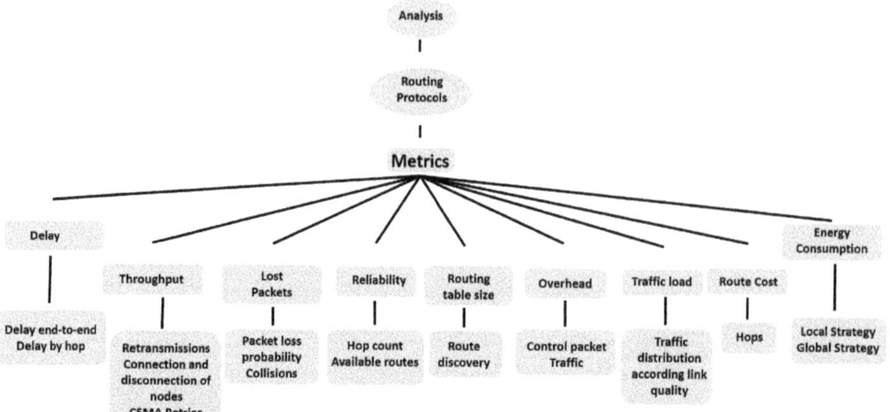

Figure 2. General performance metrics.

Sensor networks usually have similarities, so it is necessary to know some comparison criteria that allow us to decide which of them is the most suitable to implement and thus, solve our problems. The most important and useful metrics for comparing sensor networks are the following:

- *Response time:* The speed at which the data packets are sent among the different nodes.
- *Network lifetime:* The durability and resistance of the devices and, in general, of the network, must be considered if a budget is allocated or limited or both.
- *Security:* Security is vital in any field, but, in wireless networks, it is complicated to secure information. However, it is relevant to assess the risks or vulnerabilities that the network may experience.
- *Types of data that can be transmitted:* Wireless technologies such as WiFi or ZigBee, do not transmit the same types of data. It is important to select and analyze the data that it is desired to send to choose the most appropriate one. Packets that carry the information in the network are traffic packets, while the control packets are those used by the routing protocol to establish the network rules. These control packages constitute the overhead of the network, so if a network has a high overhead, there will be a higher probability of collisions.
- *Battery life:* One of the main issues in sensor networks is the energy required by the components of the masses. The life of the batteries is expected to be extended by using the application of a sleep technique to the nodes of the system.
- *Packet retransmission:* In WSNs a sensor node may need to transmit a packet multiple times to ensure delivery. This may be due to packets which have been either damaged or lost and needs to be resending or just for forwarding packets to other nodes. The indiscriminate use of packet retransmission has a negative impact on battery life.
- *Resilience:* It is the network ability to face faults or changes in its normal operation maintaining an acceptable level of service.
- *Complexity and cost:* The complexity represented by the installation of the sensor network is considered one of the most important criteria. From this, the cost can be derived, this being an agent of great relief in any project.
- *Listening to the channel retries:* According to each data link layer algorithm, the nodes listen to the channel before transmitting to know if there are collisions. After a specific number of attempts, the packet is dropped. If the number of listening tries is high, it means that the channel has high interference.

- *Packet loss:* It occurs when information packets do not arrive in the expected way. Some of the causes can be damaged hardware. Hardware capacity (bottlenecks): when some component of the network cannot assume a certain amount of traffic, thus encouraging data flow. Congestion in the network: when a device is at 100% capacity, generating a queue waiting to be discarded after a particular time if it is a long time, the difference with bottlenecks is that it can be a general problem and not just a device. Bugs in the software of network devices: When the software of the devices has errors. Other: interference or "noise" in the wireless network, proximity to other wireless devices, distance, physical elements (walls), etc. Then, the packet loss leads to the following problems: Information out of date (delay in the information received), slow loading, interruption of loads, closure of connections, incomplete information.

5. Energy

A wireless sensor network is a set of devices distributed in such a way that the sensors implemented in such network are capable of transmitting the parameters to which they are exposed, for example, temperature, pressure, humidity, etc.; in this way, a network of sensors is composed of the following elements:

- Sensors
- Radio transmitter/receiver
- CPU/memory
- Power source, usually a battery.

Regarding the energy problem, there are some aspects that directly affect node battery consumption. The term unbalanced energy depletion, shown in [42], describes a situation where the nodes that are closer to the coordinator node carry more traffic, and so they consume more energy than those nodes further away from the root node. This imbalance causes the overall energy to be distributed non-uniformly in the network, making some nodes to run out of power faster than others [43].

6. Model

There is a gateway with LoRa technology and WiFi which is the general hub of all the nodes. This gateway receives data from all devices and sends the information to a platform called the things network. The devices that are held as nodes are divided into three technologies: BLE, LoRa, and ZigBee. The LoRa modules make the direct transmission to the gateway, each LoRa node contains its own ID allowing the things network to distinguish which device is making the transmission. The BLE modules work as beacons, this allows any BLE module that is running like a beacon scanner to receive data, but in order to read the information it is necessary to decrypt the packets, this scanner contains a LoRa transmitter, this way it is possible to communicate with the main gateway and send all the information of the BLE modules. Zigbee nodes work in the same way, there is a mesh topology in which everyone communicates with each other, but there is a coordinator who creates the network, this coordinator has a LoRa module that allows communication with the gateway.

The developed network consists of four parts:

- Zigbee Nodes: A set of high-level wireless communication protocols, based on the IEEE 802.15.4 standard, i.e., communicating using the 2.4 Ghz frequency. It consists of five devices, a device is configured as coordinator, the main function of this device is to create the network in mesh topology, in this way the other four devices can be connected to the network and have communication between them, allowing an additional device withdrawn from the coordinator because the information packet passes through the other devices to the coordinator. Each of the devices contains a temperature sensor, allowing measurement of the ambient temperature in different locations within the university. The coordinator sends a signal to know which devices are connected to the network and these send their temperatures every 5 min. After obtaining all data, the coordinator sends the information through another protocol called LoRa to the gateway.

- LoRa nodes: It is a specification for low power and wide-area networks, LPWAN, designed specifically for low power consumption devices, operating in local, regional, national, or global networks. In Mexico, LoRa uses the 915 MHz frequency. The topology of LoRa is point-to-point, there is a gateway or hub and one or more nodes, the gateway oversees reading all the packets that are on that frequency. The nodes are devices that transmit small information frames to avoid a high consumption of energy. The created network consists of five different devices:

  1. Garbage sensor: This device contains a sensor called Time of Flight, which measures distance by means of infrared. The main function of this device is to measure the amount of garbage that is in a boat, in this way you can anticipate that the boat will be full, and the garbage will fall. The device makes a constant monitoring every 3 min and goes back to sleep to lower its consumption. Those monitors take a measurement of the sensor and send the value of that measurement to the gateway, after sending it back to sleep.

  2. Light Sensor: This device also using the Time of Flight sensor knows the amount of environmental light, so it is possible to know if it is day or night, if a door is open among other applications. The functionality of this device is very similar to that of garbage with the variation of the data sent to the gateway.

  3. Accelerometer sensor: The node contains an accelerometer sensor, which function is to mediate accelerations in x, y, z. This way, it is possible to know if there was displacement or some change of state of an object. This node takes a measurement every 5 min to know if there was any change in its location, if so, it sends a message to the gateway notifying with the values in x, y, z to know what its displacement was.

  4. Gyroscope Sensor: It is a mechanical device used to measure, maintain, or change the orientation in the space of an appliance or vehicle. It is essentially composed of a body with rotational symmetry that rotates around the axis of such symmetry. It informs if there was any movement on the device's own axis. The node takes a measurement every 5 min if there is no change in the sensor and sends a value. If a change occurs before the measurement is taken, the device wakes up to carry out a transmission.

  5. Environmental sensor: This node presents humidity and temperature sensors allowing knowledge of the temperature and humidity in a certain space. This node takes a sensor measurement every hour due to the little change that occurs in that period of time.

  6. Gateway: it is the device that acts as a connection interface between devices and allows sharing resources between two or more computers. The gateway used contains the LoRa and WiFi protocols. This gateway obtains all the transmitted data through the LoRa protocol, these data are then transmitted through WiFi allowing them to be found on the ThethingsNetwork platform. In this platform it is possible to display the data separately from each node. Knowing when it made its last transmission, the frequency of transmission of each device and its measurements. This device obtains all the data transmitted by the ZigBee coordinator, all the data transmitted by the LoRa nodes and the data transmitted by the BLE concentrator.

- BLE Nodes: It is an industrial specification for Wireless Personal Area Networks (WPAN), which operates on the frequency of 2.4 Ghz as well as ZigBee. There are four BLE nodes that work as beacons, it is a low-consumption device that emits a broadcast signal, i.e., they are devices that are constantly sending data. In this case they have a light sensor and they are constantly sending the value of light. In addition, there is a fifth device which works as beacon scanner, the function of this device is to receive all the data of the other nodes, decode them and send them through LoRa using another module that contains that protocol.

In Figure 3, a representative scheme of the sensors according to the wireless technology and their respective application is presented. Coordinating nodes are observed that will serve as a bridge between three of the studied technologies: WiFi, ZigBee, and Bluetooth Low Energy (BLE).

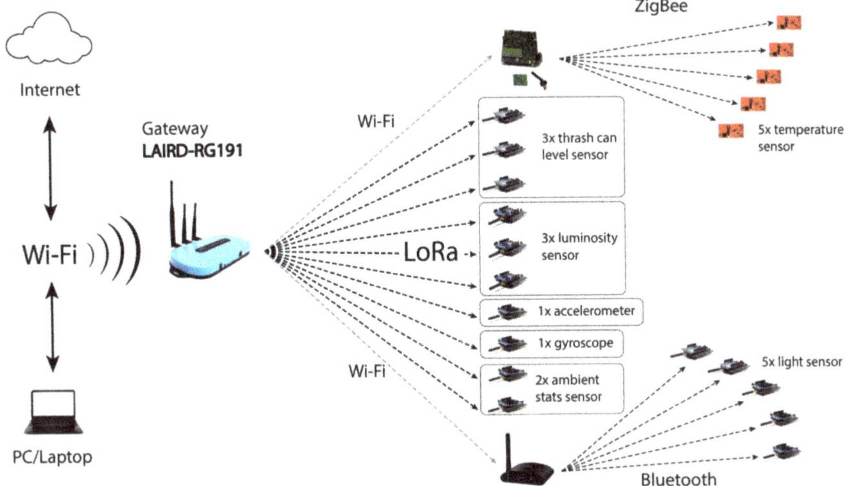

Figure 3. Network sensors under each technology.

In Figure 4, a satellite map of the university campus of the Universidad Panamericana in Guadalajara, Mexico is shown. We have an outline of the approximate positions of the main sensors, so that it serves as a basis for taking measurements. This map is useful to get an idea of the green areas, buildings, or constructions, pedestrian and vehicular routes and the approximate total area where the sensors are located.

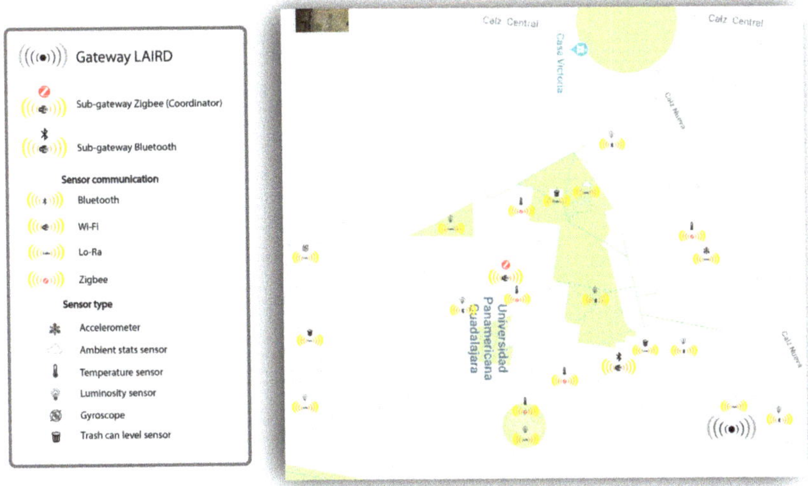

Figure 4. Network sensors position along the university campus.

The coexistence between Bluetooth and BLE in the same device is ensured by the common MAC layer. This layer also performs channel quality measurements (such as Received Signal Strength Indication (RSSI), Quality of service, and packet loss rate for Bluetooth) to update the channel map with "good" channels and eliminate those channels marked as "bad".

Scenario

Analyzing altogether the four technologies studied, we can say that BLE is used for short-range wireless connectivity with a transmission rate lower than that achieved by WiFi. BLE is characterized by transmitting small amounts of information at low frequency and allows creation of mesh networks with low energy consumption. This advantage of mesh networks takes advantage of some devices as repeaters. With respect to WiFi, there is currently an extensive infrastructure already installed that transfers data and can handle large amounts of data to provide broadband connectivity. In this technology, there is a greater loss of information, although there is greater sensitivity. In the network of the common networks of a business or academic environment, the network is saturated because there are usually many connected devices. It is an adequate standard for file transfer, but it consumes too much energy to develop IoT applications. WiFi is optimized to have many nodes connected to the same access point without causing too much packet saturation. ZigBee is a wireless technology focused on domestic and industrial applications and has significant advantages such as low consumption in complex systems, superior security, robustness, high scalability, and capacity to support many nodes. Zigbee works in the same frequency band as BLE with low data rate. One of the great advantages of ZigBee is that it presents a mechanism for a node to know when to transmit depending on the communication channel. This reduces collisions when there are multiple devices simultaneously. LoRaWAN is designed to implement wide-area networks (WANs) with specific characteristics to support mobile, bidirectional, economic, and secure communications for IoT, Machine to Machine (M2M), smart cities, and industrial applications.

The considered network is composed of a sink node and a set of homogeneous sensor nodes, which are randomly scattered in the interested area. Figure 5 is an example of one of the sensor installations on the front of a tree in a building near the entrance to the university campus. The nodes have a plastic protection box and are of easy installation.

Figure 5. Sensor in a tree in the center of a building near to the entrance of the university campus.

Figure 6 shows an example of the frames captured in the *thethingsnetwork* web application (www.thethingsnetwork.org). In this interface, through the console tab, we can see the applications that we have activated and properly configured. In each of them, an Application ID is configured to differentiate the device and a description of it. In the Data tab the configuration of the packets transmitted and received by the sensor can be found. Within the characteristics of the packets, there are the reception and transmission times, the packet count ID to know which packet is lost and the sensor data to observe the changes depending on the application. If we detail the Uplink or Downlink tabs, we find the data presented in Figure 7. Here the components of the Fields and the Metadata are described. In the Fields, we have important information about the variables that we are analyzing, such as the sensor's battery. In the Metadata, we find parameters of physical layer performance, MAC layer, and network layer, such as frequencies, modulations, interference and other parameters also related to the gateway for each technology.

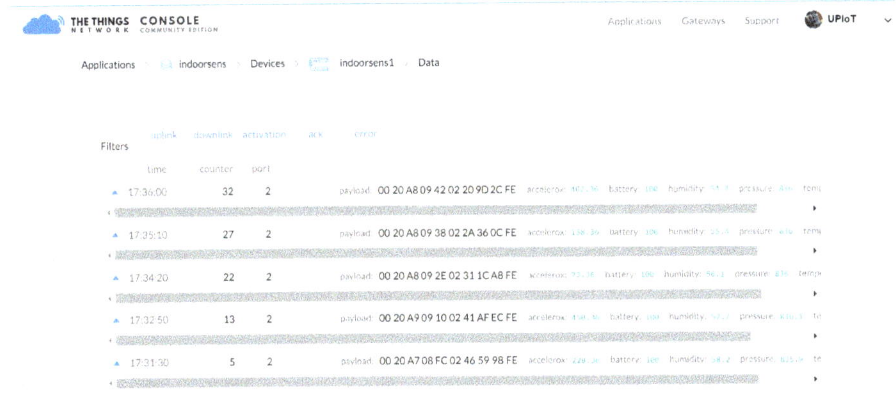

Figure 6. Data for network packets.

Figure 7. Frame of metadata for network packets.

7. Results

In the network we have implemented, the most severe aspects to be analyzed under coexistence are presented among the BLE, WiFi, Lora, and ZigBee devices, because LoRa can be configured for other frequencies. ZigBee allows the dynamic selection of channels, a scanning function goes through a list of compatible channels in search of beacon, receiver energy detection, indication of link quality. A feature called frequency agility is specified in the ZigBee standard to improve the robustness of ZigBee networks. According to this function, if an interference is detected in the current channel, a ZigBee network can move to a clear channel depending on some mechanisms. The ZigBee operation in the 2.4 GHz band is favored by the choice of the 16 available channels. The frequency agility feature facilitates the use of these additional channels. When a network is integrated for the first time, the node looks for a channel with the least noise or traffic. If additional overtime traffic appears or if there is noise, the host application looks for a better channel and moves the entire network to the new channel, allowing the network to adapt to changes in Radio Frequency RF environments. Table 1 shows the specifications of the studied wireless technologies.

Table 1. Specifications of the studied wireless technologies.

Technology	Standard	Frequency	Range	Typical Transfer Ratio
ZigBee	IEEE 802.15.4	2.4 GHz	10–100 m	250 kbps
LoRaWAN	LoRaWAN	433 MHz, 868 MHz and 915 MHz	2–15 km	0.3–50 kbps
WiFi	802.11n	2.4 GHz and 5 GHz	50 m	150–200 Mbps
BLE	IEEE 802.15.1	2.4 GHz (ISM)	50–150 m (Smart/LE)	1 Mbps (Smart/LE)

Tables 2 and 3 show the results of the different protocols under the cooperative and collaborative schemes. The metrics used together with the results should be analyzed together as they affect each other. When a node connects to the network the route discovery stage starts, in the case of ZigBee the gateway must assign an ID to the connected node. On the other hand, some protocols send control messages so that other nodes respond them, and in this way, the new node creates its neighbor tables and depending of the configuration also creates its routing tables. Hence, the more percentage of disconnected nodes the discovery process will occur more times.

Table 2. Nodes under the cooperative scheme.

Metric SD (%)	Zigbee	SD (%)	LoRaWAN		SD (%) WiFi	SD (%)	BLE	SD (%)
Transmission rate (Kbps)	250	NA	27	NA	165,000	NA	25,000	NA
Delay end-to-end (s)	2.52	1.12	3.44	1.13	2.87	1.08	3.09	1.23
Retransmissions	1.67	1.02	2.33	1.13	2.40	1.2	1.93	1.15
Channel retries	1.93	1.12	2.60	1.21	3.11	1.04	2.44	1.22
Disconnection nodes (%)	1.00	1.03	2.30	1.18	3.00	1.21	1.20	1.09
Packet loss (%)	6.00	1.01	4.00	1.06	6.0	1.10	5.00	1.02
Available routes (%)	70	1.12	80	1.13	75	1.18	90	1.02
Overhead (%)	30	1.14	35	1.08	40	1.13	30	1.21
Resilience (s)	3.00	1.24	2.50	1.12	3.00	1.26	2.20	1.22
Energy consumption (J)	2.96	1.29	3.50	1.21	5.00	1.02	4.50	1.20

Table 3. Nodes under the collaborative scheme.

Metric	Zigbee	SD (%)	LoRaWAN	SD (%)	WiFi	SD (%)	BLE	SD (%)
Transmission rate (Kbps)	250	NA	27	NA	165,000	NA	25,000	NA
Delay end-to-end (s)	2.81	1.14	3.92	1.13	3.04	1.14	3.50	1.15
Retransmissions	1.82	1.14	2.63	1.24	2.55	1.14	2.21	1.14
Channel retries	2.04	1.10	2.84	1.20	3.33	1.11	2.61	1.06
Disconnection nodes (%)	0.89	1.09	2.15	1.22	2.83	1.22	1.06	1.23
Packet loss (%)	9.00	1.13	12.0	1.08	10.0	1.19	11.00	1.20
Available routes (%)	65	1.11	74	1.21	65	1.22	83	1.14
Overhead (%)	33	1.04	40	1.08	45	1.10	38	1.20
Resilience (s)	3.20	1.20	2.80	1.11	3.50	1.21	2.90	1.22
Energy consumption (J)	3.05	1.17	3.80	1.14	5.10	1.18	4.70	1.19

Moreover, when a node tries to connect to the network, control messages are sent, therefore, the higher the percentage of disconnection of nodes, the greater the overhead. In the same way, having a greater number of packets in the network will result in more collisions and as a consequence in a greater number of retransmissions, the channel will be busier and will increase the channel retries. It is important to note that the channel retries occur when a node access the channel to verify if it is free, and if so transmit the packet. Finally, by increasing the disconnected nodes, retransmissions and channel retries the energy consumption will increase.

Otherwise, when a node is disconnected from the network some routes are canceled and new routes are created, when the node reconnects, the network enters a non-stable state, the time it takes to pass from non-stable state to a stable state is known as resilience. Therefore, the lower the resilience, the higher the energy consumption, since the network will take longer to configure and achieve new stable routes. This is directly related to the available routes, when this metric is low, there is a greater possibility of packet loss, which causes a greater number of retransmissions and finally energy consumption increases.

Tables 2 and 3 show results of performance metrics to observe the behavior of the network under the cooperative and collaborative schemes. Samples were taken from the network running for 24 h. The tests are carried out during a weekday so that the number of users and the vehicular flow is that of a normal day with activity on campus. Metrics are an average obtained every hour by all the sensors that work under each technology. The tests are carried out in an area of 300 square meters with indoor and outdoor sensors. No battery change was made for any of the sensors. The maximum number of packet retransmissions is 3, before the packet is discarded. The maximum number of retries to listen to the channel is 5, before the packet is discarded. It is well known that cross traffic affects adversely the network parameters. Thus, the scenarios experimented in this work, in practical terms, the traffic is only related to sensor network and it is expected that variation tends to be small value. A slightly variation is observed during experimentation and it can be observed that is independent of protocol. The data of the metrics studied present a standard deviation between 1 and 1.3% for the four technologies presented in each performance metrics for the cooperative and collaborative schemes. This shows that the measurements taken over 24 h can reflect a similar behavior with a small variation in the morning hours, where the traffic of people and vehicles decreases remarkably.

Specifically, Table 2 describes the results of the main performance metrics for the nodes under the cooperative scheme, i.e., when the technologies studied are in operation in the same geographical area and the priorities of the network are above the priorities of the sensor. The result of the metrics described is an average of the sensors that are under the wireless communication protocol. With respect to the end-to-end delay, ZigBee is the fastest due to its topology and message concentration. Subsequently, WiFi follows the speed of information delivery due to the greater sensitivity of the WiFi antenna and this gives an advantage, in addition to the fact that bandwidth is greater. On the other hand, the protocol with most retransmissions is WiFi, it is important to mention that a high retransmission rate increases delay and energy consumption. Additionally, retransmission as well as control packets contribute to the network overhead, being WiFi the protocol with the highest overhead, so it has a higher probability of collisions. Another important metric is resilience, in this case BLE and LoRa have the best response times, since after a failure they return to a normal operating state in a shorter time than WiFi and ZigBee.

In the collaborative scheme specified in Table 3, which consists of sharing resources and functions only if the sensor has availability, the sensor priorities are over the network priorities. In comparison with the results of Table 2, the end-to-end delay increases; however, the ZigBee remains the fastest. Following this trend, retransmissions as well as lost packets increase. Because the retransmissions increased in all the protocols, it is expected that the percentage of overhead will also increase, since mores packets are sent. Finally, it is important to mention that in critical applications where safety becomes a priority, two characteristics of the network are of high importance: resilience and availability. However, Table 3 shows that the resilience time increased with respect to the configuration of Table 2, therefore in collaborative scheme, it will take more time to achieve a normal operating state after a failure, a key factor in critical systems. On the other hand, the energy consumption directly affects the availability, most of the sensors are battery powered, i.e., the more energy they consume the faster the battery will run out. In this aspect, the collaborative scheme is better, since it consumes less energy, being ZigBee the protocol with the lowest consumption.

These Figures 8 and 9 show random measurements (11 for each wireless technology). These figures are intended to show the intensity of the received signal. In this way, we can see under which scheme we can determine if a signal is sufficient to establish a wireless connection when other technologies and diverse environmental situations are present. Then, each of these measurements helps us to know the strength with which the devices listen or could hear the signal at that specific point. Likewise, we observe that for both schemes, WiFi is the technology with higher RSSI levels (in dBm), which shows that there are areas of very low signal coverage, while the technology that exhibits the best levels (very good coverage) presents is LoRa, possibly because its operating frequency is different and faces fewer collisions. The position of each measurement was performed in a uniformly distributed manner throughout the area of the university campus, where the sensors were located. For the cooperative scheme shown in Figure 8, the average RSSI value in dBm for BLE is -90.36, for ZigBee it is -85.36, for WiFi it is -99.09 and for LoRa it is -68.36. For the collaborative scheme shown in Figure 9, the average RSSI value in dBm for BLE is -92.73, for ZigBee it is -87.91, for WiFi it is -100.73 and for LoRa it is -71.82. We can then state that both schemes differ between 8 and 10% with the cooperative scheme being less interfering. It is important to bear in mind that the best transmission signals are observed in values close to -75 dBm. This makes sense considering that the LoRa technology is the one with the best coverage and, therefore, will present the least number of collisions. The one that presents worse coverage is WiFi and is understandable due to the large number of devices connected to campus networks, it is logical that the characteristics of a campus present a smaller scale than the characteristics of a city. Of the three technologies that are in the same frequency band, ZigBee has better signal coverage (between good and medium coverage). These figures help to better understand Tables 2 and 3 due to the metrics related to the interference of the channel such as: collisions, packets retransmissions, resilience, among others.

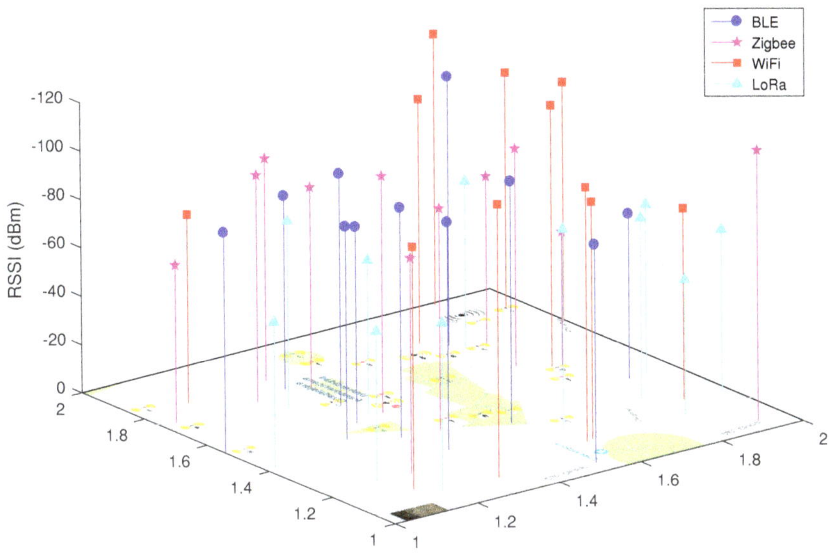

Figure 8. Received Signal Strength Indication (RSSI) values in dBm under cooperation scheme for each technology.

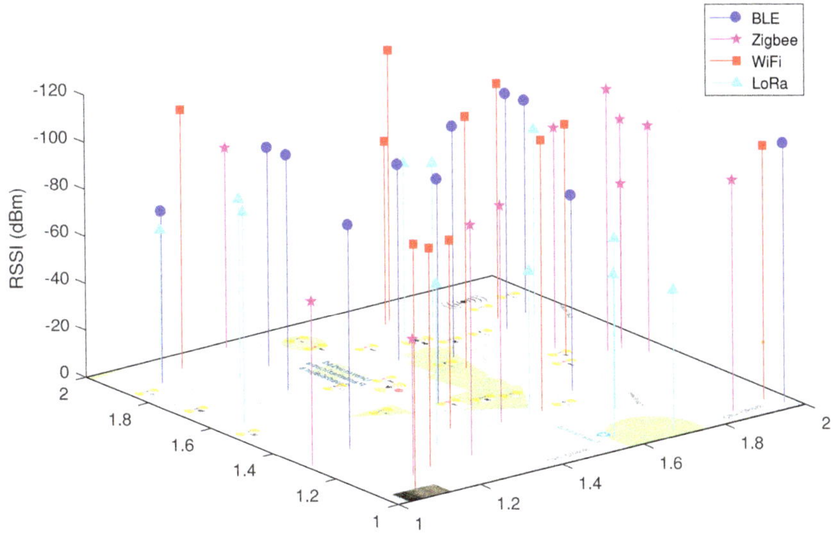

Figure 9. RSSI values in dBm under collaboration scheme for each technology.

As mentioned above, one of the key metrics in any sensor network is energy consumption. Figures 10 and 11 show a summary of the energy consumption for each protocol under the cooperative and collaborative schemes.

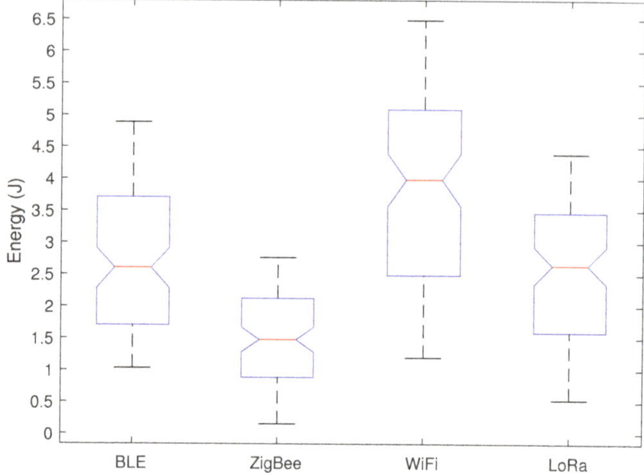

Figure 10. Energy in Joules under cooperation scheme for each technology.

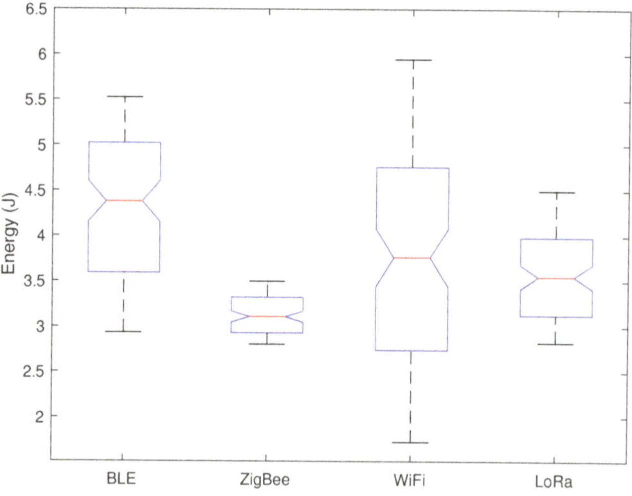

Figure 11. Energy in Joules under collaboration scheme for each technology.

Figure 10 shows that under the cooperative scheme the protocol with the lowest energy consumption is ZigBee, where the minimum consumption value is 0.2 Joules, but most of the sensors consume between 0.8 and 2.1 Joules, with the median being 1.5 Joules. Zigbee consumes about half of the LoRa protocol, since the latter has an average value of 2.4 Joules; however, 50% of the sensors consume between 2.4 and 4.3 Joules. On the other hand, BLE and WiFi have very scattered data, reaching to consume values as low as 1.1 Joules up to 6.8 Joules as in the case of WiFi, with the median value of 3.2 Joules for BLE and 4.8 Joules for WiFi. In this way, it is clear that the protocol with the best consumption is ZigBee, additionally it has less dispersed data so that a more exact battery life calculation can be made, otherwise in WiFi the scattered data would cause different battery life time for each sensor.

By contrast, Figure 11 shows the protocols under the collaborative scheme; however, the results are very similar to those in Figure 10, with the ZigBee protocol having the lowest energy consumption. It is important to mention that in a comparison between schemes, the cooperative has a lower energy consumption in all protocols except for WiFi, in which values are very similar. The biggest difference is the dispersion of the data, for the ZigBee protocol the range in the cooperative scheme is 3.6 Joules and in the collaborative scheme is 0.7 Joules, something similar occurs for both LoRa and BLE. This low dispersion indicates that all the sensors consume practically the same energy (at least in ZigBee), so that the lifetime of their batteries will be very similar, being able to program a more efficient maintenance of the network.

8. Conclusions

The main motivation of this work is to contrast low-consumption wireless technologies applied to IoT that are characterized by transmitting small amounts of information in a reliable and flexible way, consume little battery in devices, and have great scalability in the communications system.

Concerning data, we can control the information management of applications in our computers such as photos, videos, mails, etc. But for IoT, it does not work in the same way. It captures information at every moment it considers necessary, instead of capturing it when requested. People usually become subjects for data collecting, instead of users of IoT services, and most cases they are not aware of it. Because it is not easy for people to know when sensors are activated. The privacy stack framework bridge of today's Internet of Things between IoT and user, starts with awareness. It concerns on how IoT services might open communication channels to users and subjects. The IoT protocol work has not gone into privacy data standardization, in other words, the bridge between privacy and public status is minimal. Second, the inference part proposes users to be conscious of the constantly grow of inferences, because data and IoT learning techniques rise their capabilities every day. Inferences helps users to understand what IoT devices learn about them and helps the system to improve privacy with a natural language to understand which are our privacy preferences.

Author Contributions: The methodology was proposed by C.D.-V.-S., the investigation and the formal analysis work were done by L.J.V. and C.D.-V.-S., the validation work was done by R.V., the data analysis work was finished by J.-C.L.-P. and L.R.-D., and the original draft was finished by L.J.V., C.D.-V.-S. and R.V.

Funding: This research received no external funding.

Conflicts of Interest: The authors declare no conflict of interest.

References

1. Yigitel, M.A.; Incel, O.D.; Ersoy, C. QoS-aware MAC protocols for wireless sensor networks: A survey. *Comput. Netw.* **2011**, *55*, 1982–2004. [CrossRef]
2. Aswale, P.; Shukla, A.; Bharati, P.; Bharambe, S.; Palve, S. An Overview of Internet of Things: Architecture, Protocols and Challenges. In *Information and Communication Technology for Intelligent Systems*; Springer: Berlin, Germany, 2019; pp. 299–308.
3. Silva, J.D.C.; Rodrigues, J.J.; Al-Muhtadi, J.; Rabêlo, R.A.; Furtado, V. Management Platforms and Protocols for Internet of Things: A Survey. *Sensors* **2019**, *19*, 676. [CrossRef] [PubMed]
4. Sikandar, A.; Kumar, S.; Singh, P.; Tyagi, M.K.; Kumar, D. Energy Efficient Transmission in the Presence of Interference for Wireless Sensor Networks. In *International Conference on Application of Computing and Communication Technologies*; Springer: Berlin, Germany, 2018; pp. 55–64.
5. Liang, J. Collaborative Mechanism of Enhanced Coexistence of Collocated Wireless Networks. US Patent 7099671B2, 29 August 2006.

6. Sisinni, E.; Saifullah, A.; Han, S.; Jennehag, U.; Gidlund, M. Industrial internet of things: Challenges, opportunities, and directions. *IEEE Trans. Ind. Inform.* **2018**, *14*, 4724–4734. [CrossRef]

7. Luo, L. Data Acquisition and Analysis of Smart Campus Based on Wireless Sensor. *Wirel. Pers. Commun.* **2018**, *102*, 2897–2911. [CrossRef]

8. Alavi, A.H.; Buttlar, W.G. An overview of smartphone technology for citizen-centered, real-time and scalable civil infrastructure monitoring. *Future Gener. Comput. Syst.* **2019**, *93*, 651–672. [CrossRef]

9. Del Esposte, A.D.M.; Santana, E.F.Z.; Kanashiro, L.; Costa, F.M.; Braghetto, K.R.; Lago, N.; Kon, F. Design and evaluation of a scalable smart city software platform with large-scale simulations. *Future Gener. Comput. Syst.* **2019**, *93*, 427–441. [CrossRef]

10. Fortino, G.; Russo, W.; Savaglio, C.; Shen, W.; Zhou, M. Agent-oriented cooperative smart objects: From IoT system design to implementation. *IEEE Trans. Syst. Man Cybern. Syst.* **2018**, *48*, 1949–1956. [CrossRef]

11. Lghoul, R.; Abid, M.R.; Khallaayoun, A.; Bourhnane, S.; Zine-Dine, K.; Elkamoun, N.; Khaidar, M.; Bakhouya, M.; Benhaddou, D. Towards a real-world university campus micro-grid. In Proceedings of the 2018 International Conference on Smart Energy Systems and Technologies (SEST), Sevilla, Spain, 10–12 September 2018 .

12. Lee, I.; Lee, K. The Internet of Things (IoT): Applications, investments, and challenges for enterprises. *Bus. Horiz.* **2015**, *58*, 431–440. [CrossRef]

13. Dhanda, S.S.; Singh, B.; Jindal, P. Wireless technologies in IoT: Research challenges. In *Engineering Vibration, Communication and Information Processing*; Lecture Notes in Electrical Engineering; Springer: Singapore, 2019; Volume 478, pp. 229–239.

14. Yang, D.; Xu, Y.; Gidlund, M. Wireless coexistence between IEEE 802.11- and IEEE 802.15.4-based networks: A survey. *Int. J. Distrib. Sens. Netw.* **2011**, *7*, 912152. [CrossRef]

15. Bauwens, J.; Jooris, B.; Giannoulis, S.; Jabandžić, I.; Moerman, I.; De Poorter, E. Portability, compatibility and reuse of MAC protocols across different IoT radio platforms. *Ad Hoc Netw.* **2019**, *86*, 144–153. [CrossRef]

16. Mahmoud, M.S.; Mohamad, A.A.H. A Study of Efficient Power Consumption Wireless Communication Techniques/Modules for Internet of Things (IoT) Applications. *Adv. Internet Things* **2016**, *6*, 19–29. [CrossRef]

17. Chiwewe, T.M.; Mbuya, C.F.; Hancke, G.P. Using Cognitive radio for interference-resistant Industrial wireless sensor networks: An overview. *IEEE Trans. Ind. Inform.* **2015**, *11*, 1466–1481. [CrossRef]

18. Grimaldi, S.; Mahmood, A.; Gidlund, M.; Alves, M. An SVM-based method for classification of external interference in industrial wireless sensor and actuator networks. *J. Sens. Actuator Netw.* **2017**, *6*, 9. [CrossRef]

19. Soffker, P.; Block, D.; Wiebusch, N.; Meier, U. Resource Allocation for a Wireless Coexistence Management System based on Reinforcement Learning. In Proceedings of the 2018 IEEE 23rd International Conference on Emerging Technologies and Factory Automation (ETFA), Turin, Italy, 4–7 September 2018; pp. 1101–1104.

20. Adeyemi, O.J.; Popoola, S.I.; Atayero, A.A.; Afolayan, D.G.; Ariyo, M.; Adetiba, E. Exploration of daily Internet data traffic generated in a smart university campus. *Data Brief* **2018**, *20*, 30–52. [CrossRef] [PubMed]

21. Lee, J.; Su, Y.; Shen, C. A Comparative Study of Wireless Protocols: Bluetooth, UWB, ZigBee, and Wi-Fi. In Proceedings of the 33rd Annual Conference of the IEEE Industrial Electronics Society (IECON 2007), Taipei, Taiwan, 5–8 November 2007; pp. 46–51. [CrossRef]

22. Saad, C.; Mostafa, B.; Cheikh, E.A.; Abderrahmane, H. Comparative Performance Analysis of Wireless Communication Protocols for Intelligent Sensors and Their Applications. *Int. J. Adv. Comput. Sci. Appl.* **2014**, *5*. [CrossRef]

23. Naidu, G.A.; Kumar, J. Wireless Protocols: Wi-Fi SON, Bluetooth, ZigBee, Z-Wave, and Wi-Fi. In *Innovations in Electronics and Communication Engineering*; Saini, H.S., Singh, R.K., Kumar, G., Rather, G., Santhi, K., Eds.; Springer: Singapore, 2019; pp. 229–239.

24. Mostafaei, H.; Montieri, A.; Persico, V.; Pescapé, A. A sleep scheduling approach based on learning automata for WSN partialcoverage. *J. Netw. Comput. Appl.* **2017**, *80*, 67–78. [CrossRef]

25. Drossos, N.; Mavrommati, I.; Kameas, A. Towards ubiquitous computing applications composed from functionally autonomous hybrid artifacts. In *The Disappearing Computer*; Springer: Berlin, Germany, 2007; pp. 161–181.

26. Ochiai, H.; Mitran, P.; Poor, H.V.; Tarokh, V. Collaborative beamforming for distributed wireless ad hoc sensor networks. *IEEE Trans. Signal Process.* **2005**, *53*, 4110–4124. [CrossRef]

27. Chen, H.; Zhai, C.; Li, Y.; Vucetic, B. Cooperative strategies for wireless-powered communications: An overview. *IEEE Wirel. Commun.* **2018**, *25*, 112–119. [CrossRef]

28. El-Mougy, A.; Ibnkahla, M.; Hattab, G.; Ejaz, W. Reconfigurable wireless networks. *Proc. IEEE* **2015**, *103*, 1125–1158. [CrossRef]

29. Choi, Y.S.; Shirani-Mehr, H. Simultaneous transmission and reception: Algorithm, design and system level performance. *IEEE Trans. Wirel. Commun.* **2013**, *12*, 5992–6010. [CrossRef]

30. Ping, L.; Liu, L.; Wu, K.; Leung, W.K. Interleave division multiple-access. *IEEE Trans. Wirel. Commun.* **2006**, *5*, 938–947. [CrossRef]

31. Velazquez-Gutierrez, J.M. Reception of Multiple Users in Reconfigurable Wireless Networks. Ph.D. Thesis, Tecnologico de Monterrey, Monterrey, Mexico, 2018.

32. Hong, J.P.; Choi, W.; Rao, B.D. Sparsity controlled random multiple access with compressed sensing. *IEEE Trans. Wirel. Commun.* **2014**, *14*, 998–1010. [CrossRef]

33. Fang, F.; Zhang, H.; Cheng, J.; Leung, V.C. Energy-efficient resource allocation for downlink non-orthogonal multiple access network. *IEEE Trans. Commun.* **2016**, *64*, 3722–3732. [CrossRef]

34. Khedr, M.E.; Zaghloul, M.S.; El-Desouky, M.I. Wireless Adhoc Multi Access Networks Optimization Using OSPF Routing Protocol Based On Cisco Devices. *Int. J. Comput. Netw. Commun.* **2015**, *7*, 59. [CrossRef]

35. Wang, D.; Gao, X.; You, X.; Han, B. Channel Estimation Algorithms for Broadband MIMO-OFDM Systems. *Acta Electron. Sin.* **2005**, *33*, 1254.

36. Mahfoudh, S.; Minet, P. Survey of energy efficient strategies in wireless ad hoc and sensor networks. In Proceedings of the Seventh International Conference on Networking (icn 2008), Cancun, Mexico, 13–18 April 2008; pp. 1–7.

37. Raya, M.; Hubaux, J.P. Securing vehicular ad hoc networks. *J. Comput. Secur.* **2007**, *15*, 39–68. [CrossRef]

38. Xiao, Y.; Chen, H.; Yang, S.; Lin, Y.B.; Du, D.Z. *Wireless Network Security*; National Institute of Standards and Technology: Gaithersburg, MD, USA, 2009.

39. Du, D.Z.; Ko, K.I. *Theory of Computational Complexity*; John Wiley & Sons: Hoboken, NJ, USA, 2011; Volume 58.

40. Alazzawi, L.; Elkateeb, A.; others. Performance evaluation of the WSN routing protocols scalability. *J. Comput. Syst. Netw. Commun.* **2009**, *2008*, 481046.

41. Fauzia, S.; Fatima, K. Performance evaluation of AODV routing protocol for free space optical mobile Ad-Hoc networks. In *The International Symposium on Intelligent Systems Technologies and Applications*; Advances in Intelligent Systems and Computing; Springer: Cham, Switzerland, 2018; Volume 683, pp. 74–83.

42. Babayo, A.A.; Anisi, M.H.; Ali, I. A Review on energy management schemes in energy harvesting wireless sensor networks. *Renew. Sustain. Energy Rev.* **2017**, *76*, 1176–1184. [CrossRef]

43. Tuah, N.; Ismail, M.; Haron, A.R. Energy consumption and lifetime analysis for heterogeneous Wireless Sensor Network. In Proceedings of the 2013 IEEE TENCON Spring Conference, Sydney, Australia, 17–19 April 2013; pp. 188–193.

MDPI

St. Alban-Anlage 66

4052 Basel

Switzerland

Tel. +41 61 683 77 34

Fax +41 61 302 89 18

www.mdpi.com

Energies Editorial Office

E-mail: energies@mdpi.com

www.mdpi.com/journal/energies